Simply Modelling Global Warming
Second Edition

: Global Warming Handbook (2)

By Jim Wiles

Extensively reviewed following the issue of the 2021 IPCC Physical Science Basis (AR6) and extended to include the effect of the Oceans and Energy Imbalance.

Paperback

Dedication

To Kella,

To Bill,
who once told me that you can't model industrial plant to better than 10%

Disclaimers

This book has been produced in good faith and the author has tried to ensure that content is accurate. No responsibility is taken for any errors or omissions. No liability is taken for any use of its content.

Contextual links are provided to aid understanding. The author takes no responsibility for and cannot guarantee the accuracy or suitability of information on sites or documents that are not under his management; nor should the inclusion of a hyperlink be taken to mean that he endorses the sites to which it points.

Copyright © 2020 & 2022 Jim Wiles. All rights reserved.

Other Books in the series

Global Warming Handbook - by Jim Wiles

This is the first book in the series. It looks at a complicated subject and strives to make it simple, dealing with it gradually and without drama. – All in a short book.

<u>Simulation Made Easy – by Jim Wiles</u>

How to produce **Dynamic** Models the Easy Way, with step-by-step graded examples - from basic mathematical functions to the simple physics of a pendulum, and ultimately to Engineering systems, **Global Warming** and even COVID-19.

Contents

What's in this book?..1
1 - Global Warming..3
2 - How Does our World Warm?..7
3 - Solar Input..9
    Solar Input Calculation..9
4 - The Atmospheric Window..13
    The Atmospheric Window Calculation....................................15
5 - Atmospheric Transfer..19
    Atmospheric Transfer Calculation...20
6 - Working out our Temperature Rise.......................................25
7 - The Effect of Carbon Dioxide..33
    Making Use of the IPCC CO_2 Forcing Relationship..............33
8 - The Effect of Other Greenhouse Gasses..............................37
    Including Other Greenhouse Gasses......................................37
9 – Aerosols..43
    Including Aerosols in the Model..43
10 - Temperature Feedback...47
    Water Vapour Feedback..47
    Evaporative Feedback...49
    Increased Atmospheric Capture of Sunlight...........................52
    Increased Absorption of Window Radiation...........................53
    Atmospheric Feedback Roundup..53
    Increased Atmospheric Water Vapour....................................54
11 – Ozone and Other Forcings..57
    Other Forcings...57
    Surface Albedo Feedback...57
    Ozone...58

12 – The Oceans and Energy Balance 61
13 – Model Prediction Current and Future 67
   Comparison with Measurements..**67**
   Future Prediction **68**
14 – Uncertainties 72
15 - Global Warming Spreadsheet 77
16 - Final Thoughts 79
Other Books in the Series 81
The Author 82
Acknowledgements 83
Appendix 1 - Absorption by CO_2 85
Appendix 2 - The IPCC Forcing Relationship... 88
Appendix 3 - IPCC Forcing Relationship Validity.. 91
Appendix 4 – Changing the Model Reference Date 94
Appendix 5 – Evaporative Feedback – the Balance of Evaporation 97
Appendix 6 - Atmospheric Heat Gain Due to Increased Water Vapour 107
   Change in Absorption of Sunlight **107**

   Water Vapour Absorption Effects on Thermal Window Radiation......**109**

Appendix 7 - Calculating Radiative Forcing... 111
Appendix 8 – Guide to the Spreadsheet 113
   Using the Spreadsheet **113**

What's in this book?

In this book you will find the details of the science and the component parts that make up a Global Warming model - established, step by step. You may be surprised to learn that it uses little more than high school physics.

This book follows on from the 'Global Warming Handbook', which explains global warming with the aim of making the science accessible. To support this, a simple model was used to illustrate the effect of each warming process.

Here, you will find not only the details of the model and the explanations of the science but also a book structured to allow you to choose the depth that suits your needs. To aid this, I have deliberately created the model as a spreadsheet. If you have access to Microsoft Excel, you will be able to download this spreadsheet from the *http://www.Earth2hot.com* website.

At another level, if you are familiar with this area, you may also be surprised to find a different approach, where the subject is reduced to fundamentals where topics such as water vapour feedback can be realised from some simple physics and maths and the effect of the Ocean Deeps explained and simply quantified.

You might have many reasons for reading this book. You may simply want to gain an understanding of the modelling; you may want to create your own model, you may want to use it as a teaching aid, or you may want to try out the spreadsheet and use the book to guide you. My goal is that all of these can be satisfied in this book.

Using a model gives a greater appreciation of what is happening in a physical system – in this case, our planet - not only providing insights but enabling 'what if' questions to be asked and tested. It allows you to make your own judgements.

Jim Wiles

Website: *http://www.Earth2hot.com*,
Jim Wiles *Earth2hot@gmail.com*

Abbreviations and Organisations Noted in this Book

IPCC	Intergovernmental Panel on Climate Change (UN)
AR6	The sixth report published by the IPCC. (The Physical Science Basis - 2021)
AR5	The fifth report published by the IPCC. (The Physical Science Basis - 2013)
AR4	The fourth report published by the IPCC (2013. (The Physical Science Basis - 2007)
NOAA	National Oceanic and Atmosphere Administration (USA)
AOML	NOAA's Atlantic Oceanographic and Meteorological Laboratory
NASA	National Aeronautics and Space Administration (USA)
CO_2	Carbon Dioxide
TAO	Tropical Atmosphere Ocean (Monitoring Buoys)
ppm	parts per million – used to measure the concentration of carbon dioxide
WMGHG	Well Mixed Greenhouse Gases

1 - Global Warming

This short book does not repeat the content of the Global Warming Handbook: but, to stand alone, parts of the initial sections are included with enhanced content. If you have read this first book, you can choose to skip the initial few pages.

Global Warming occurs because of an increase in heat within our climate systems. This occurs when greenhouse gasses trap the flow of heat that would otherwise be radiated directly into space. It is most easily recognised by the rise in temperature of the surface air immediately above the continents and oceans. We will touch on climate processes in the next sections, but first let's look at changes in global temperature.

Global temperature change is assessed from thousands of measurements. Originally these were from weather stations dotted around the globe and from sea temperatures, available for shipping purposes. These measurements go back a long way, with the standardisation of temperature measurements set up in 1872. Much of the work in recent times has concentrated on checking and unifying this data as well as adding to it. More recently, sea temperatures are derived from measurements that now strive to investigate ocean temperatures at various depths.

In many publications, the term 'surface temperature' is often used. It is important to realise that this relates to a combination of ocean and land air temperatures and not the temperature of the ground we stand on. On land, it is based on temperatures in the shade measured at a standard height of a meter and a half or so above the ground. Climate scientists use these temperature measurements to work out what is normal for the mid-twentieth century. Using this as a reference point enabled them to highlight what is unusual and they found that temperatures before this reference point are generally lower and after this point are higher.

Figure 1.1 uses information provided by NASA and shows differences between measured temperatures and the twentieth-century norm. They called these 'anomalies', meaning something unusual. This is why temperatures for early dates are shown as negative (cooler than the norm), and those for later dates are positive (hotter than the norm).

To put things simply, it is best to ignore what is positive or negative and view the chart overall as showing a trend in temperatures from 1880 to current times. The main idea of this type of graph is to highlight the change in temperature from pre-industrial times up to the present.

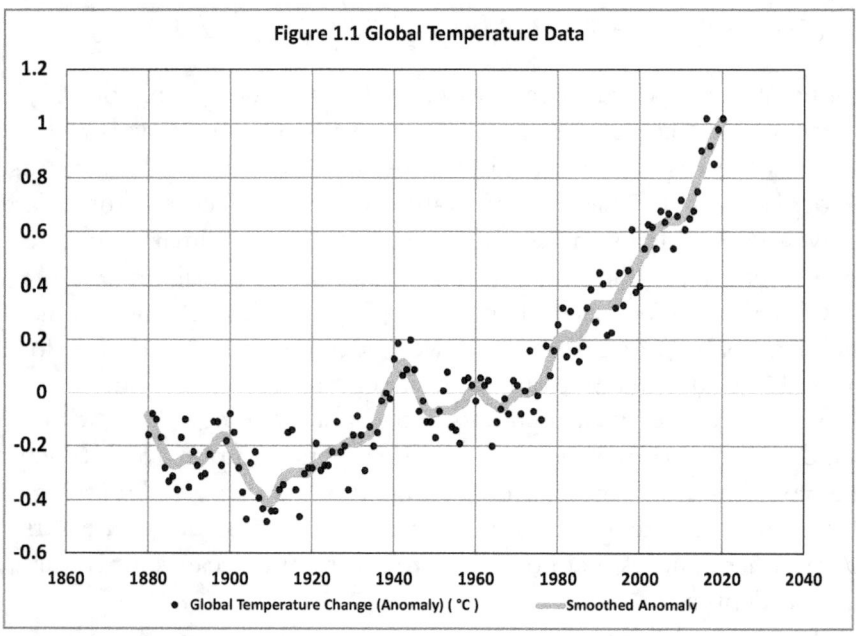

The first thing to note is that 'year by year', our global temperature has changed by small and erratic amounts. However, the overall difference between 1880 and 2020 is statistically significant. It is a real temperature increase. Our concern is that we do not want it to increase because, as small as it is, it can affect our climate.

Notes:

Graphs of 'temperature anomalies' can seem odd, with their early negative values. However, they are used here because this is the standard way this information is displayed elsewhere.

The United Nations has sponsored a body called the Intergovernmental Panel on Climate Change (IPCC). Its objective is to provide the world with a clear scientific view of the current state of climate change as well as its potential environmental and socio-economic impacts. It does this by producing reports for worldwide distribution. You can find these on the internet, and throughout this book, you will find them frequently referenced. You should note that whilst this book uses and compares data in IPCC reports, it also uses other sources. Whilst its aim is not to parrot IPCC documents, it does respect the integrity of the scientists undertaking the supporting work.

Further Information:

IPCC, AR4, **Turn to page 100,** Section 1.3.2 Global Surface Temperature.
https://www.ipcc.ch/site/assets/uploads/2018/03/ar4-wg1-chapter1.pdf

The website CarbonBrief outlines how several organisations make their independent assessment of our global temperature change.
https://www.carbonbrief.org/explainer-how-do-scientists-measure-global-temperature

Figure 1.2 A Climate Monitoring Buoy under Repair

2 - How Does our World Warm?

Energy comes from our sun in the form of radiation (sunlight). It strikes our earth and heats land, sea, and atmosphere. Through a multitude of processes, this heat is distributed by our oceans and our atmosphere, eventually rising to the upper atmosphere - where it radiates back into space.

A balance is reached where 'what comes in must also go out'. If this did not happen, we would get hotter and hotter, and our planet would cease to support life. To achieve a balance, natural processes act to ensure the temperature of the planet is raised by exactly enough to push the incoming heat back into space. The result is a stable surface temperature. Into this mix there is also an important leakage path which plays a vital role.

If we impair any aspect of this global heat transfer, then the earth's surface warms, and its temperature rises. It's a bit like putting on an extra jumper. This extra jumper impedes the heat transfer from our bodies, and we warm up.

The three principal processes involved in creating our climate balance are:

1. **Solar Input**
   (sunlight)
2. **Atmospheric Heat Transfer**
   (Results in heat radiating from the atmosphere into space)
3. **Atmospheric Window Radiation**
   (Is heat radiating directly from the surface into space)

Figure 2.1 (on the next page) shows the overall way in which this happens, and the next few sections explain each individual process and the basis on which they can be modelled.

The sun heats the earth and its temperature rises. When earth's temperature rises by exactly enough, the heat going out balances the heat coming in.
At this point we have a stable temperature.

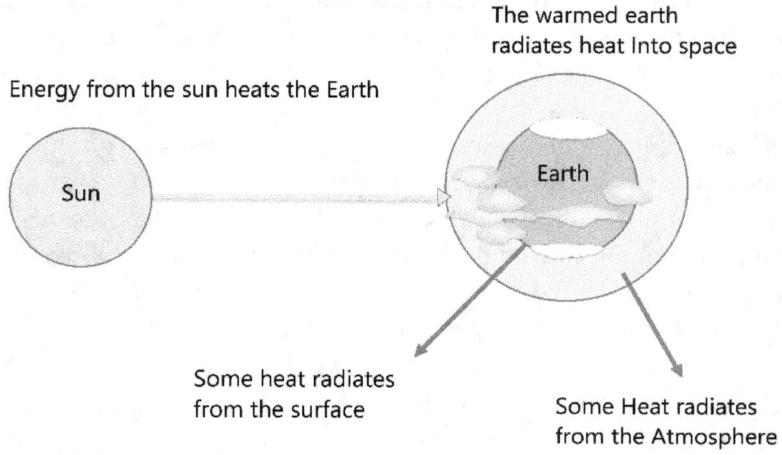

The warmed earth radiates heat Into space

Energy from the sun heats the Earth

Sun

Earth

Some heat radiates from the surface

Some Heat radiates from the Atmosphere

Figure 2.1 Earths Heat Balance

3 - Solar Input

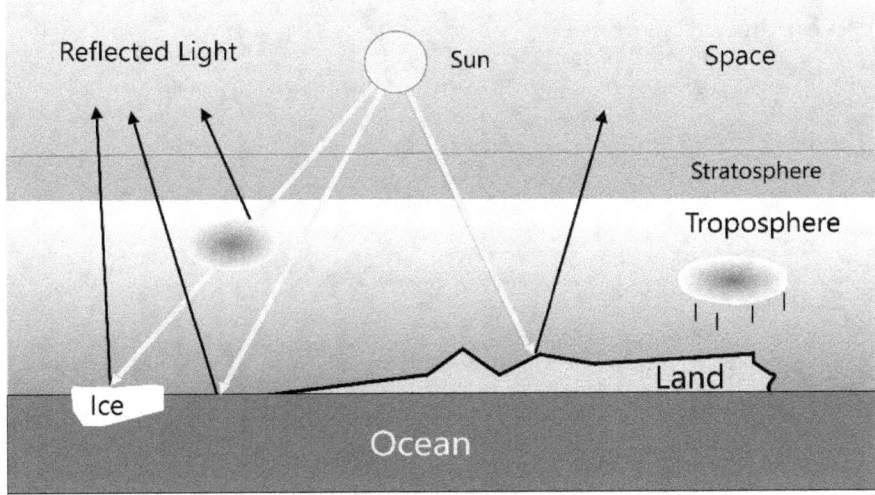

Figure 3.1   Solar Input

Energy reaches the earth from the sun in the form of radiation (sunlight). It travels through space to reach us and during the day, it bathes half of our globe in light. This light has a spectrum, a range of wavelengths that is characteristic of the sun's temperature. This is termed Short Wave radiation by climatologists. It is the light we can see with our eyes - and includes ultraviolet, which we cannot see. Some sunlight is immediately reflected by clouds and the earth's surface and this returns immediately to space. The amount of solar energy that reaches our planet, after allowing for reflections, will cause it to warm.

Solar Input Calculation

This is the easiest process to evaluate. The power coming directly from the sun is well known. This is called the Solar Constant and is given in Watts per square metre at the top of the atmosphere. However, not all this solar energy warms the planet. We need to deduct the amount reflected back into space by clouds, snow, and the earth itself. This reflection or Albedo factor is readily available and is well known from satellite measurements. All that remains is knowledge of the earth's radius so that we can calculate the area that captures solar radiation. I chose to extend this radius slightly to include half the bulk of the atmosphere - a height of 5km. Therefore, we can obtain all the values we need for this calculation.

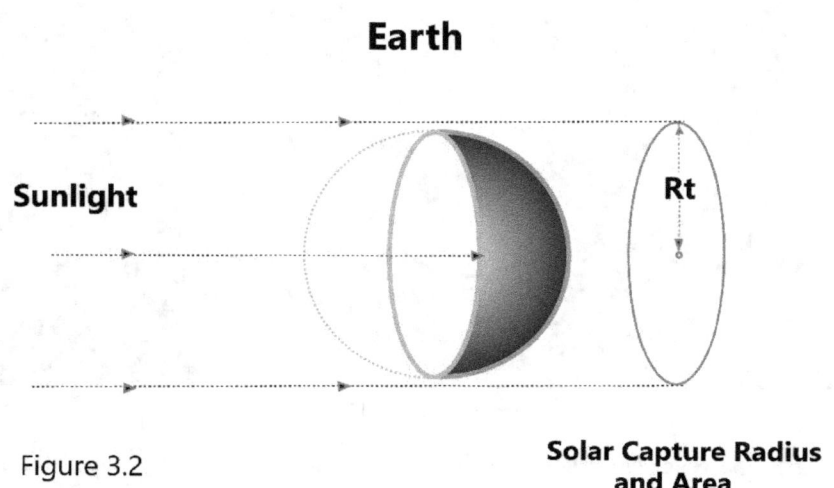

Figure 3.2 — Solar Capture Radius and Area

Our basic formula for the solar input power becomes:

SolarPower = Sc x (1-albedo) x CaptureArea (W)

Equation 1

Where:
SolarPower is the effective solar input power (W)
Sc is the solar constant (1366 W/m^2)
'**albedo**' is the proportion of reflected sunlight (0.3)
'**CaptureArea**' is $\pi \times Rt^2$ or 1.2772×10^{14} m^2
Where **Rt** is earth's solar capture radius (6377km)

Here, we are calculating the power arriving from the sun, measured in Watts. From figure 3.2 you can see how this is caught by the cross section of the earth. Only one side of the earth faces the sun at any point in time.

The value of this is simply 1.221×10^{17} W - a lot of power. In practical terms, it is better to put the formula into a spreadsheet where the calculated values go into a larger calculation.

One reason to do this is that we may need to alter the albedo (the fraction of sunlight reflected from the earth). For example, the Arctic ice cap has been found to be reducing. As snow-covered ice is a particularly good reflector of sunlight, this will make a change to the earth's albedo. You can then see what happens to global temperatures as you take this into account.

Notes:

*Here, we are calculating the power arriving from the sun, measured in **Watts**. In IPCC documents, they often refer to the power received in **Watts per square metre**, which is spread over the surface area of half of a globe (termed **Flux**). Do not confuse these two values – they are both easily calculated.*

Further Information:
NASA, Measuring Earth's Albedo
https://earthobservatory.nasa.gov/images/84499/measuring-earths-albedo

4 - The Atmospheric Window

This is our world's safety valve. Atmospheric water vapour does not allow the energy arriving from the sun to pass directly back to space: it captures it.

However, water vapour does not capture the entire spectrum. This means there is a route for some of the heat radiated from the earth's surface to travel directly into space. See Figure 4.1.

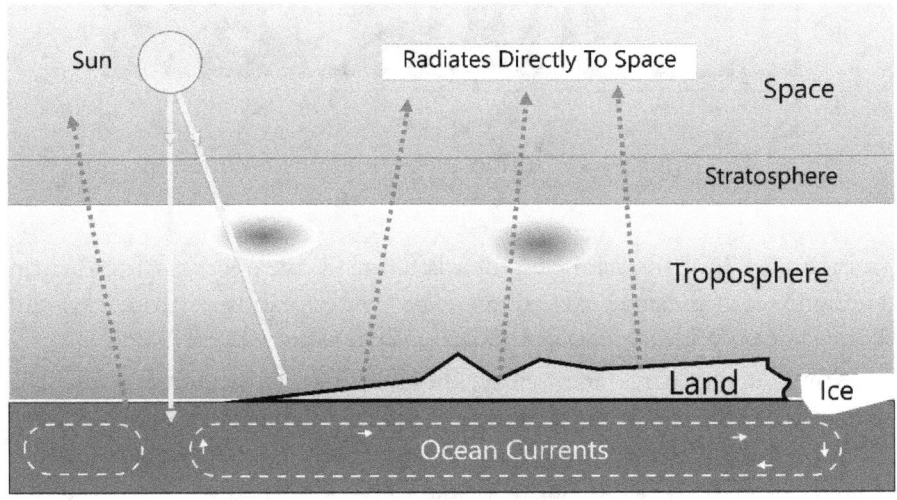

Figure 4.1 Atmospheric Window

Only a tiny proportion of heat can use this route, with over 95% directly absorbed by water vapour. However, this bypass route through the earth's atmosphere is enough to help regulate the earth's surface temperature, and within this book, I call it the 'thermal Atmospheric Window'.

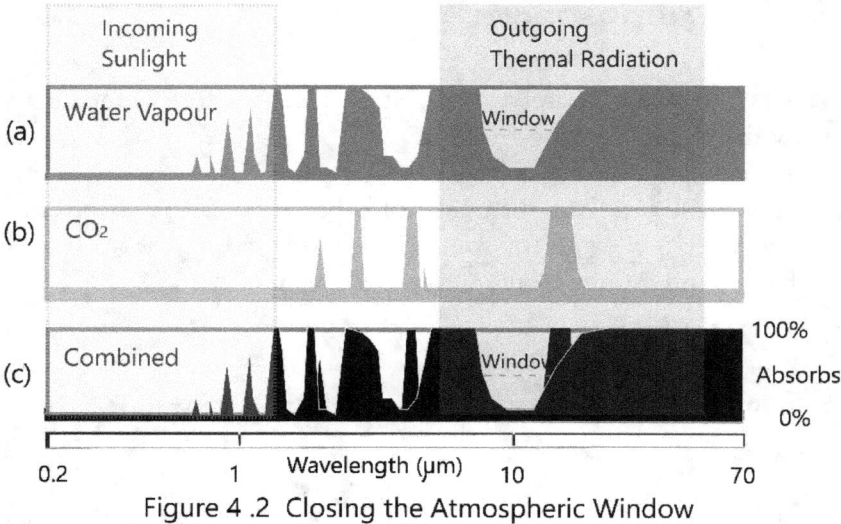

Figure 4.2 Closing the Atmospheric Window

Figure 4.2(a) shows how the different wavelengths of radiation (sunlight and thermal) are treated by water vapour. Note that an Atmospheric Window allows thermal radiation to escape.

Now we can introduce carbon dioxide. Just like water vapour, it is selective in the spectrum of thermal radiation that it captures. Figure 4.2(b) shows the absorption characteristics of CO_2, and you can see that one of the CO_2 absorption peaks lies within the water vapour atmospheric window. In essence, it blocks up our safety valve and the results are shown in (Figure 4.2(c)) This is why a small amount of CO_2 can have a very significant impact upon global temperatures. It is this that currently makes CO_2 the largest driver of global warming.

Despite this, we do need a certain level of carbon dioxide to get just the right temperature balance. Overall, CO_2 is an immense benefit. Without CO_2, our planet's average surface temperature would drop below freezing. Perhaps more importantly, if we did not have CO2, plant photosynthesis would stop completely, taking with it our source of oxygen. All life on earth would then cease. So, CO_2 is not in itself bad – it is vital.

In pre-industrial times (before 1890), our CO_2 level was about 288 parts per million. By January 2021, our CO_2 level reached 414 parts per million and this is what is relevant to global warming.

The Atmospheric Window Calculation

As explained above, the amount of energy escaping directly into space from the earth's surface depends upon absorption by greenhouse gasses.

In the absence of these gasses, energy will be radiated according to the physical laws of radiation.

$$\text{Radiated Power} = \sigma \cdot A_e \cdot T_s^4$$

Where **Ts** is the surface temperature

A_e is the entire earth's surface area

σ is Stefan's constant

When we introduce greenhouse gasses, this becomes:

$$\text{Radiated Power} = T_{gh} \cdot \sigma \cdot A_e \cdot T_s^4$$

Where **Tgh** is the transmission factor allowing for the absorption by greenhouse gasses. We can relate this to absorption by these gasses:

$$T_{gh} = (1 - \text{Absorption})$$

This can be split into two parts:

$$T_{gh} = [\,1 - a_{1ref} - a_2\,]$$

Where a_{1ref} is a reference level of water vapour absorption (including, for the moment, other greenhouse gasses) and a_2 is absorption by CO_2, which varies with CO_2 concentration. Therefore, the power radiated via the thermal atmospheric window at a given surface temperature becomes:

$$\text{WinPower} = [1 - a_{1ref} - a_2] \cdot \sigma \cdot A_e \cdot T_s^4$$

This leaves us with the issue of working out how a_2 varies with CO_2 level.

IPPC documentation explains that the relationship for the absorption of thermal radiation from Earth's surface, by CO_2, is a process of diminishing returns. The more CO_2 we add to our atmosphere, the less effect it has. They explain that the absorption process can be represented by a logarithmic relationship. In turn, they make use of a logarithmic relationship to describe the change in thermal radiative power with CO_2. Because it has not been possible to locate a published IPCC relationship specifically for absorption, we need to synthesise our own. I have used a simple relationship for CO_2 absorption of the general form:

$$a_2 = a_{2ref} \cdot (1 + \ln(C/C_0))$$

Where, C_0 is a CO_2 reference level in ppm and C is some other concentration value where we want to evaluate absorption. Looking at this relationship, a_2 must equal the reference value when C is set to the reference concentration of C_0 – because, in this instance, '$\ln(C/C_0)$' is 0.

The change in absorption when CO_2 level changes to some level C, is '$a_{2ref} \cdot \ln(C/C0)$'. In other words, the change in absorption is directly proportional to the logarithmic change in CO_2 level. Summarising:

The transmission factor is,

$$Tgh = [1 - a_{1ref} - a_{2ref} \cdot (1 + \ln(C/C_0))] \quad \text{Equation 2}$$

and the thermal window radiated power is,

$$WinPower = Tgh \cdot \sigma \cdot A_e \cdot T_s^4 \quad \ldots\ldots \text{Equation 3}$$

Where T_s is the global average surface temperature arising from the completed model. Climate scientists chose to express this radiative power loss in terms of Watts per square metre. This is referred to as Radiative Flux:

$$Winflux = Tgh \cdot \sigma \cdot T_s^4$$

Now at a given reference temperature Ts_{ref} we can determine Tgh_{ref} if we know **Winflux**. In contemporary times, NASA evaluate Winflux as 40Wm^{-2}. I have chosen this to relate to a reference date of 2005 when the average global surface air temperature was measured as 287.83 K and measured CO_2 was 378.91ppm. This yielded a value for Tgh_{ref} of 0.10278.

The final piece in this analysis is the decision to express the CO_2 concentration at this reference level as a percentage of all greenhouse gases ($a_{2ref} + a_{1ref}$), from which a_{2ref} and a_{1ref} can be simply calculated.

Absorption reference examples for 1.53% CO_2, at 2005, are:

$$a_{1ref} = PercentCO_2 \, (1 - Tgh_{ref})/100$$

$$(a_{1ref} = 0.01373)$$

$$a_{2ref} = (100 - PercentCO_2)(1 - Tgh_{ref})/100$$

(a_{2ref} = 0.88349)

CO_2 absorption at some other CO_2 level is then directly calculable from **Equation 2 above.** This analysis in terms of percentage at some reference level is central to the way in which this model works.

Finally, Winpower for any modelled **Ts** value can calculated using **equation 3.** We now have a complete set of expressions to enable us to calculate the effect of CO_2 on the radiation of energy from the earth's surface.

Notes:

- *The analysis above covers only the effect of CO_2. This will be developed further in Section 7 (to consider a percentage figure embracing spectrographic analysis noted in IPCC reports) and in Section 8 (to include other greenhouse gasses).*
- *Whilst a static reference level of water vapour absorption is used above, in later sections you will see how changes in atmospheric water vapour content are dealt with.*
- *The value for Winflux is one of several values stated / updated by NASA. I have used a value of 40 W/m^2 for consistency with my initial work.*

5 - Atmospheric Transfer

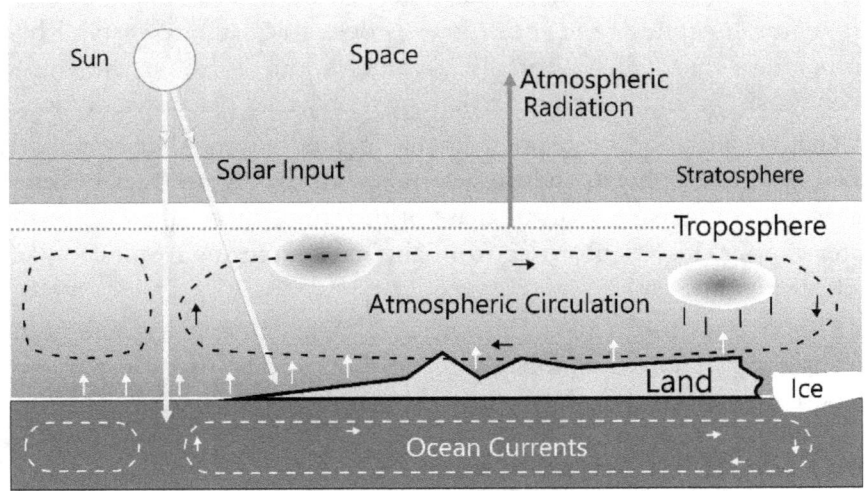

Figure 5.1 Atmospheric Transfer

Sunlight passes through our atmosphere, and part of it is absorbed by the atmosphere. However, the bulk is absorbed by ocean and land surfaces. These are warmed and give off thermal radiation, called infrared or what climatologists call 'Long-Wave' radiation. Thermal radiation is the type of radiation that we can all feel as warmth but cannot see. Unlike the Short-Wave radiation, coming directly from the sun, this infrared radiation is readily absorbed by water vapour in the atmosphere and most of it does not immediately escape. Incidentally, the bulk of our atmosphere, nitrogen and oxygen, does not impede the flow of either incoming or outgoing radiation.

Clearly, this thermal energy must escape, or our world would rapidly get hotter and hotter. Crucially, as noted above, water vapour does not allow this heat to pass directly back to space – it captures it. Not only does it capture it, but it re-radiates it. To leave our planet this captured and re-radiated heat is shared, with land and sea and with the rest of earth's atmosphere. Ultimately, it mixes and joins the earth's atmospheric circulation systems to be finally lost at a high altitude. This altitude is where our atmosphere thins sufficiently to allow outward radiation. Essentially, the atmosphere radiates heat into space at some point high above us. Figure 5.1 shows the overall picture. Importantly, the surface temperature must rise to drive this overall heat transfer process; a process where evaporation of water plays a key part. We are extremely lucky that our world is a blue planet and that water is abundant because this natural capture and heat loss process is vital to establish a temperature that will support life. Water vapour is by far the largest greenhouse gas and our planet would be a very cold place without it.

That part of our atmosphere that contains our weather systems is called the 'troposphere', and this extends about 10km above the earth's surface. It contains 75% of the mass of the atmosphere and 99% of the water vapour. Its height varies from about 17km in the tropics to 6km in polar regions.

Although I have called this process 'Atmospheric Transfer', the oceans are a vital part of global heat distribution. The name was chosen because heat escapes from the upper atmosphere into space.

Atmospheric Transfer Calculation

This is a massively complex area, but as far as global warming is concerned, it is just a process by which heat returns to space, giving us a certain surface temperature. However, as far as climate change and our lives are concerned, it is the interaction of air and land and sea, of their many currents and flows; of frontal systems, of ice sheets, of storms and typhoons and hurricanes, of deserts glaciers and tropical forests; of human impact. In one sense you cannot work out temperature change without taking on board all of these. That is what we will do – and to achieve this we have one great ally.

Our climate systems (our heat transport systems) have one unifying aim: they strive to reach an equilibrium. If our climate systems are capable of achieving stability under a range of conditions – and ancient history shows that they are - then we have our simple answer. All that I mention above is driven by this one equilibrium-seeking goal. Our climate mixes the incoming heat in a hurly-burly. At one point in time, a given area on our planet is hot only to go cold, as day is followed by night, as winter follows summer, as rainy seasons give way to drought, as rivers catch water and send it to the oceans, as the ocean currents drive poleward only to return, as CO_2 increases. So, our planet is an enormous mixing heat engine.

The important thing is that at any point in time we now have available a global temperature assessment (see section 1). However, whilst seeking equilibrium, the parts of the globe are never in equilibrium, but global equilibrium is represented by our global surface temperature **average**. Some areas are hotter, and some are colder. All will interact with sea, air, and land, but when we average them correctly, we have a measure of our equilibrium. Without realising it, we are defining our equilibrium state.

This means that on a global-wide analysis, seeking to rationalise worldwide average temperatures, we can simplify heat transfer to only one term. The only thing that such an approach is incapable of giving is the issue of time scales or the intervening dynamics. On the other hand, if we are seeking answers over decades then we should have a viable estimate of global temperature rise. When the model described in this book is complete, you can judge the outcome for yourselves.

Basic physics assures us that, unaided, heat energy can only flow from a warmer to a colder body. In this case the warmer body is characterised by its global surface temperature, an average covering thousands of measurements over the entire globe. I am calling this **Ts**. At some height in the troposphere, this heat is radiated into space. This will occur at a notional, or rather an effective temperature **Te**, which will occur at an effective altitude. There is nothing new about the concept of heat being radiated in the troposphere, and this is embedded in IPCC thinking. Our first step in calculating our overall heat transfer co-efficient is a simple heat transfer equation:

$$AtmosPower = K2 \cdot (Ts - Te) \text{ Watts}$$

Equation 4

Where K2 is our overall atmospheric heat transfer coefficient.

In the previous section we found that we can independently work out the power transmitted through the thermal atmospheric window (WinPower). Given the availability of a calculated WinPower.

$$AtmosPower = (SolarPower - WinPower) = K2 \cdot (Ts - Te)$$

At a specific measured temperature reference value of Ts, we can work out K2:

$$K2 = (SolarPower - WinpowerRef) / (Ts_{ref} - Te_{ref})$$

Equation 5

However, we need a value for **Te$_{ref}$**. Fortunately, we know that at a temperature **Te,** heat must be radiated according to the laws of radiation.

AtmosPower = σ · A$_t$ · Te$_{ref}^4$

and

Te$_{ref}$ = (AtmosPower / σ · A$_t$)$^{0.25}$

Where 'A$_t$' is the surface area at a point on the troposphere.

Also, as:

AtmosPower = (SolarPower − WinPower)... from above:

Te = [(SolarPower − Winpower)/ (σ A$_t$)]$^{0.25}$

Equation 6

at any simulated set of conditions.

At this point, we have all the information we need to evaluate Equation(5) and calculate **K2**, our overall heat transfer coefficient. Importantly, we also have the means to solve Equation (4) (making use of Equation (6), which relies only on Winpower from the previous section). In summary, we now have a basis for evaluating power radiated into space via the atmospheric transport system.

I suggest that we use a fixed value of **K2**, over the range of conditions from 1880 to 2019. This is only possible because global warming temperature changes are numerically small (compared to absolute temperatures of around 280 K). Whilst **K2** was evaluated using 2005 conditions, my basic checks indicated that if it were evaluated using measured conditions from 1970 to 2018, we would see less than 2% variation. The calculation of a specific value for **K2** requires some more information, for example, a value for the percentage of CO_2 as a proportion of the overall greenhouse gas effect. We will return to this shortly.

I must conclude with one important point. Please do not confuse my statement above about numerically small Global Temperature Changes – this relates to the calculation - their impact on climate may not be trivial.

Having read sections 3,4 and 5 you are now in a position to model global temperature response to changes in CO_2 level. Let's see how that is done.

Notes:

- *Climatologists do not consider water vapour to be the cause of recent warming as they calculate that our industrial contribution to water vapour is insignificant. However, we will see later that it can act to amplify the effect of greenhouse gasses.*

Further Information:
Atmospheric Circulation Videos – well worth viewing. Apologies for any YouTube adverts - if they occur – press "skip adverts" to move past.
https://www.metoffice.gov.uk/weather/learn-about/weather/atmosphere/global-circulation-patterns

Wikipedia – Gulf Stream – an example of ocean current heat transfer.
https://en.m.wikipedia.org/wiki/gulf stream

<*continued overleaf*>

Figure 5.2 Cumulonimbus Thunder Cloud

The top of the troposphere is called the Tropopause. We can all see it from time to time if we know where to look. It is the point where the highest clouds cease to rise, and you can see this most clearly by the flat top of a thundercloud. This storm is even spilling into the stratosphere.
Image NOAA/AOML/Hurricane Research Division

6 - Working out our Temperature Rise

By now, you should have a general picture of how the greenhouse gasses, water vapour and carbon dioxide (CO_2) work in regulating our planet's temperature. In addition, you will also have encountered something of the physics and the equations that determine the planet's energy balance. This takes us a long way in creating a model. There are other greenhouse gasses, notably methane and nitrous oxide, and other processes at work, but we will leave these for later. So how do we combine the outcome from the previous sections (all those equations) so that can estimate the change in surface temperature with changes in CO_2?

The answer is to create a simple numerical global warming model that includes the three processes explained above and use a spreadsheet to balance them and calculate the average global surface temperature for a given CO_2 value.

This is really a two-step process. First, we need to decide the reference conditions that we want to apply to our simulation and calculate the static parameters explained in Sections 3, 4, and 5. In my spreadsheet model, I do this once on a separate spreadsheet page.

Figure 6.1 shows this diagrammatically, highlighting the reference conditions and what they produce.

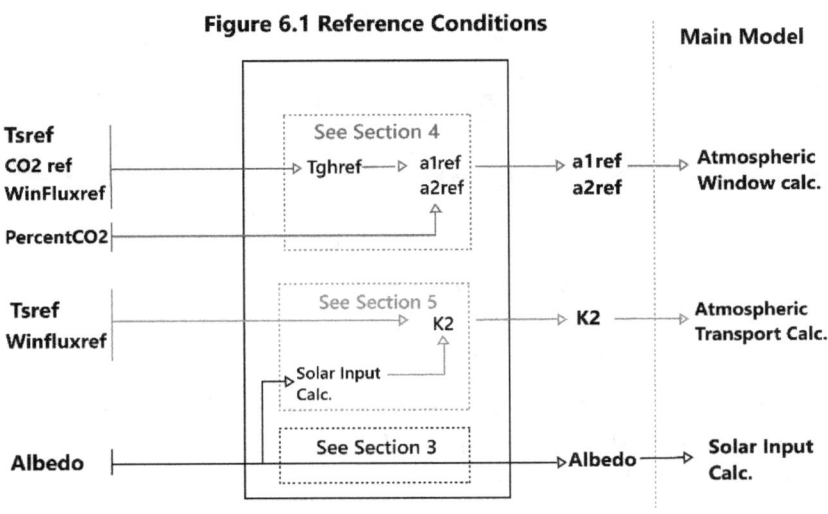

The conditions that I have used are:

Reference	Value
Tsref	287.83 K (measured value 2005)
CO_2ref	378.9 ppm (measured value 2005)
WinfluxRef	40.0 W/m² (contemporary value NASA)
Percent CO_2	Discussed in Sections 7 and 8
Albedo	0.3 (dimensionless)

Note that these values are always used as our reference conditions. However, it is often more useful to translate them to equivalent 1890 conditions, and this process is described in Appendix 4. I am conscious that readers will have differing purposes in using this book. Some will want to construct their own model, and they will need this background.

CO_2 absorption - Available Reference Conditions - these apply to all Worksheets				Thermal	Relative	
		Ts	CO_2	WinFlux	Absorption	Albedo
		(K)	(ppm)	(W/m²)	(CO_2 %)	
Preset Conditions						
1	○ Try 3.62% 1890 ref	286.80	292.46	42.48	3.620	0.3000
2	○ Try 2.77% 1890 ref	286.80	292.46	41.85	2.770	0.3000
3	○ IPCC CO2 Only 1890 Reference	286.80	292.46	41.10	1.530	0.3000
4	● All WMGHG 1890 Reference	286.80	292.46	41.64	2.665	0.3000
Initial Reference examples						
5	○ IPCC CO2 only 2005 Ref	287.83	378.90	40.00	1.530	0.3000
6	○ All WMGHG 2005 Ref	287.83	378.90	40.00	2.371	0.3000
User Values						
7	○ User Values 1					
8	○ User Values 2					
9	○ User Values 3					
You can edit the User Values and then select them						

However, many may simply download my spreadsheet, which already provides a control panel to use pre-set reference conditions. Note that CO_2 levels will shortly be discussed.

Having covered reference conditions, our next step is to explain how the model itself works This is a straightforward energy balance model, and the way that it works is shown in Figure 6.2

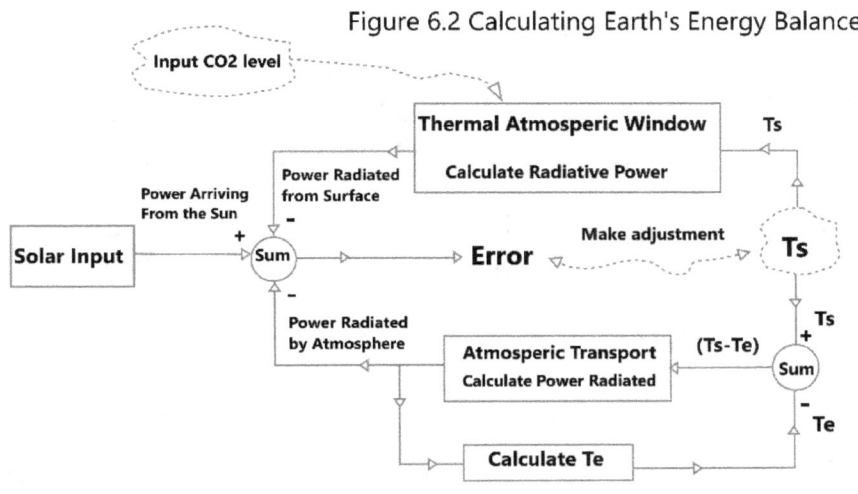

Figure 6.2 Calculating Earth's Energy Balance

In Figure 6.2, we have provided an input of a certain CO_2 level and an initial guess has been made for Ts, our air surface temperature. The model then calculates the heat flow for the three components that influence global temperature. As the heat coming from the sun must exactly equal the heat radiated back to space, the outgoing heat is subtracted from the incoming heat. If this is zero, then we have our heat balance. To make it zero we can adjust Ts. Each time we adjust Ts we can see whether the error is positive or negative and make a corresponding adjustment to Ts. If we do this correctly, then each time we make a correction we should find that the error reduces. When it is small enough, we have found our heat balance and our surface temperature Ts for the CO_2 level we have used.

Now, it would be tedious to do this by hand, but most spreadsheets have an iterative capability where the computer makes the adjustments for you. When using Excel on my laptop, this is an instantaneous calculation, taking a fraction of a second. This iterative approach comes into its own when we eventually need to include feedback processes.

Each box in figure 6.2 contains the calculations, previously explained in section 3, 4, and 5.

Energy Balance Calculation	Source
Solar Input	Equation 1
Thermal Atmospheric Window	Equations 2 and 3
Atmospheric Radiation	Equation 4
Calculate Te	Equation 5

When preparing a spreadsheet of this type, it is important that the calculation progresses either horizontally or vertically, with one of the initial cells containing a value for Ts and the final cell containing the correction to be fed back. This final cell will contain a correction of the form:

Ts = Ts + (Error) x (ScaleFactor)

The value of ScaleFactor is important. If you make it too big, then the corrections are too big, and the solution bounces around, taking longer to converge and probably heading for infinity. On the other hand, if you make ScaleFactor too small then the solution may again take a long time and give an erroneous answer due to something called 'rounding error' (essentially, we approach the resolving power of the spreadsheet).

For my model, which conducts its calculations in terms of 'power', not 'flux', I used a value for ScaleFactor of 1.022×10^{-16} (K / W). This is deliberately precise because it is the time scale factor that would be used in a dynamic model (a more sophisticated time-varying model) of a similar type. You can use an alternative value if you wish. In simple terms you need to have a ScaleFactor value small enough to reduce corrective changes in Ts to a sensibly small level, bearing in mind that the solar input power is of the order of 10^{17} watts. (but see note 2 at the end)

Now that we have a model that we can run - what percentage value for CO_2 should we try out on our model?

Well, first, we should try something simple - zero per cent CO_2 and no water vapour. The outcome is a model prediction of a mean surface temperature of minus18.2°C. This frozen planet is exactly as expected by climatologists. They often express it in degrees Kelvin as 255K. So, our model predicts that without CO_2, and water vapour, we would all freeze to death.

Now let's try 3.62%. This was a figure I often saw quoted on the internet, particularly on sites critical of climate change. I was unable to find its source until I recently found it attributed to a respected physicist, John Tyndall, who looked at heat absorption by gasses in 1860, but again, I could not find independent confirmation of this.

Figure 6.3 shows a graph of the actual measured temperature changes since 1880, drawn with a broad line. You should also notice that it only uses temperatures from a small number of dates to clarify what is happening.

Figure 6.3 also shows us what happens if our model uses 3.62% for the reference absorption of CO_2. This is the thin line marked with circles. We can immediately see that it is a reasonable match with early measurements but diverges with later dates. To be fair, the original assessment (if attributed correctly) was done in 1860.

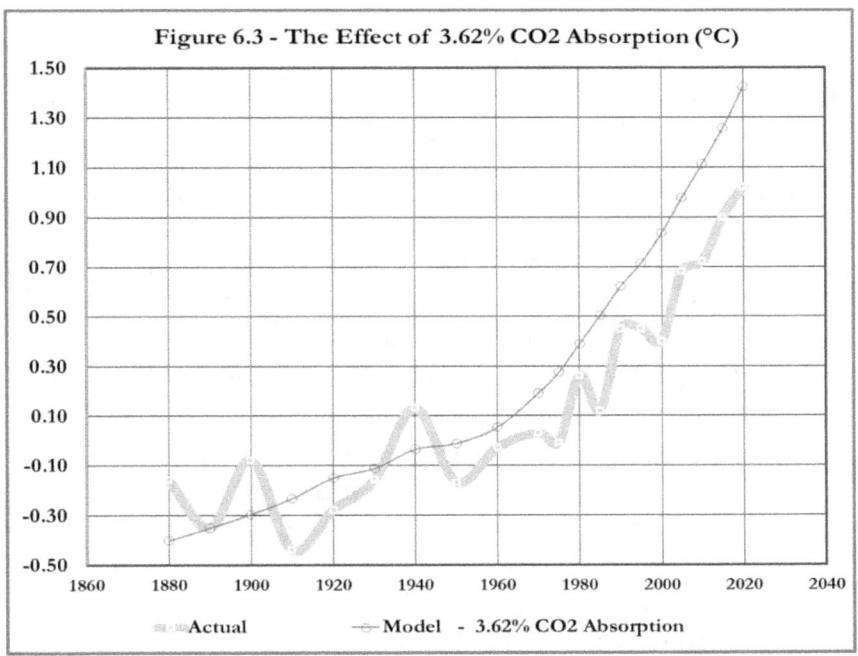

We can also make our own guess at 2.77% related to 2005. In Figure 6.4 we can immediately see a good match over the entire period.

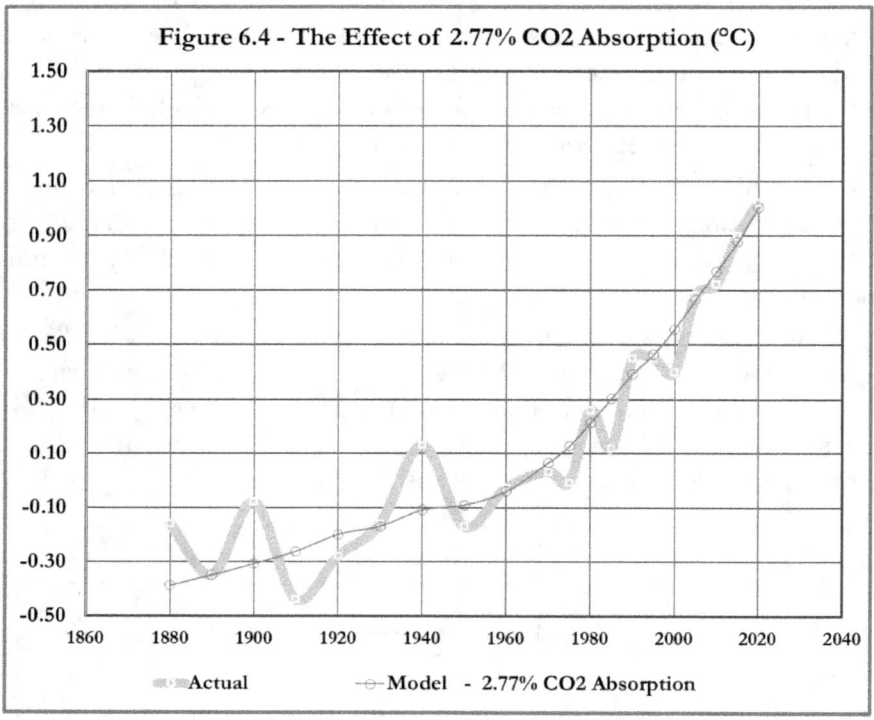

In all honesty, please don't be impressed with the two values we used. All we did was try out an assessment, probably from 1860, then pulled another percentage out of the hat and tried that. Whilst we may be left with a feeling that we are only playing around, we did achieve one thing:

- **Using the model, we have seen that when we inject a small percentage of CO_2 into our atmosphere, we get a significant change in temperature.**

Unfortunately, our best guess 'model line' is only valid for predicting temperatures from 1880 to 2020.

We really want to go forward in time. We want to know where global warming will take us, and we cannot do this without a **cause-and-effect** relationship between CO_2 level and absorption. We need to have a scientific theory. This would allow us to look ahead in time – and the next section provides this.

Figure 6.5 **Summit Supercomputer**

Notes:

- *1/ Climate simulations can take months to run. That is why they need supercomputers like Summit – the world's fastest supercomputer, which occupies a space of two tennis courts and weighs 340 tons.*
- *Photograph - Carlos Jones/ Oak Ridge National Lab*

- *2/ If you have a background in numerical techniques, you will have noted that the Excel spreadsheet model explained here has the same form as an integrator: and the model is, in effect, a type of dynamic model. Note that a truly dynamic version of this model has been run on more sophisticated software and gives similar results.*

- *3/ Some details of dynamic Global Warming models are provided in:*

-

Simulation Made Easy – by Jim Wiles
Available from Amazon

7 - The Effect of Carbon Dioxide

Moving on from the previous section, we need a scientifically established value for absorption in order to predict future temperature changes.

First, let's once more go back in time. In 1900. A notable scientist, Knut Angstrom, published laboratory experiments that showed that CO_2 was not a significant greenhouse gas. He concluded that, at the levels of CO_2 present at that time, the absorption of CO_2 would be saturated. Any increases in CO_2 would have no additional effect on the absorption of infrared and no effect on global temperatures.

Analysis in later times showed that CO_2 could continue to absorb infrared via a small part of its absorption spectrum. However, this residual absorption ability gradually reduced as CO_2 levels increased. Unfortunately for us, this remaining absorption still lay within the water vapour thermal atmospheric window and it could cause global warming.

Interest in this area increased, and everyone from astronomers to satellite systems engineers needed precise absorption data for atmospheric gasses. The effect of CO_2 and many other gasses has now been evaluated and reviewed many times. The scientists supporting IPCC made use of this and converted it to something called changes in 'Radiative Forcing' (power changes per unit area), sometimes called RF.

This means that we can take the IPCC Radiative Forcing relationship and convert it back to absorption so that the model will work on the basic physics of the issue.

Making Use of the IPCC CO_2 Forcing Relationship

Radiative Forcing is defined as the instantaneous power change (per square metre) at the top of the troposphere, induced by a change in some variable, e.g., CO_2. For us, it is a very useful parameter because it tells us about the change in power for a given change in CO2 in isolation from all other changes that may arise from other causes. It relates directly back to spectrographic absorption.

An IPCC reference provides a relationship between changes in Forcing (ΔF) for a change in carbon dioxide (ppm) as:

$$\Delta F = 5.35 \ln(C/C_0) \text{ W/m}^2$$

Where C_0 is the reference CO_2 (ppm) value at a reference date or period, and C is the CO_2 concentration at some other time.

It is straightforward to convert this into the percentage absorption of CO_2 with respect to all greenhouse gasses (including water vapour). The way in which this is calculated is explained in Appendix 2. From this, we learn that CO_2 absorption based upon IPCC forcing is 1.53% using 2005 as a reference year (1.53% is the proportion of CO_2 absorption with respect to all greenhouse gasses -but dominated by water vapour).

The outcome is shown in Figure 7.1. You should note that to produce this graph, the model reference date is shifted to 1890. I found this a convenient early date when temperature measurements were available.

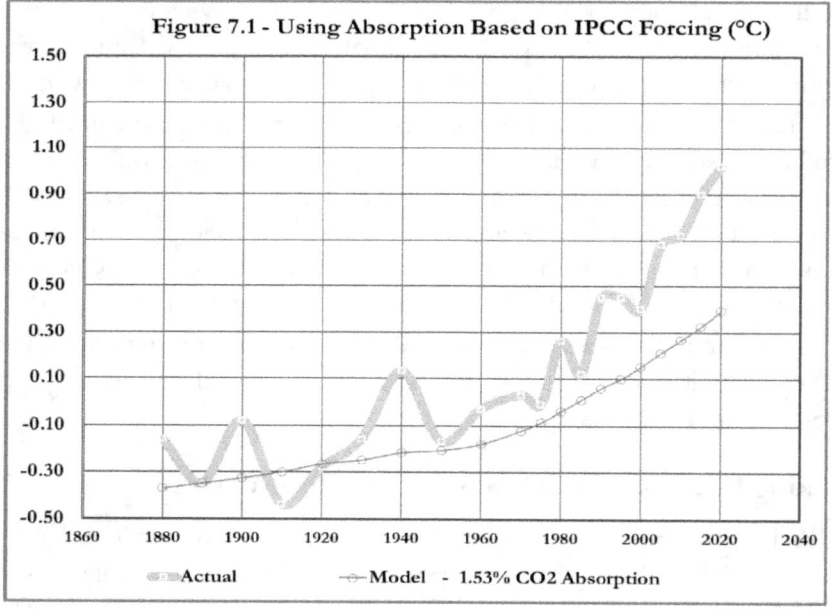

Moving the reference point is a straightforward process and it is outlined in **Appendix 4**.

Returning to Figure 7.1, it looks a bit of a disappointment. The modelled temperatures (the line marked by black circles) are clearly much lower than those actually measured. A cross-check on the model response for 2019 shows that its Radiative Forcing (from the IPCC reference date of 1750) is 2.1 Wm^{-2} and well within the uncertainty for the IPCC AR6 forcing figure of 2.16 ±0.259 Wm^{-2}. The latest IPCC figure (AR6) is based upon a variant of RF, termed Effective Radiative Forcing (ERF), which also allows for other changes, but for greenhouse gasses, we are told that ERF and RF values are essentially identical. Nevertheless, in constructing this model, I have always attempted to get back to the basic RF. If you read IPPC documents and references, you will see this spectrographic assessment referred to as the 'kernel'.

As we now seem to have alignment between IPCC assessed spectral data and our model, we need to move on because an important part of our model appears to be missing?

Notes:

The logarithmic relationship between CO_2 and Radiative Forcing relationship used here is a 'simplified expression' and was reported in 1998. A negligible numerical difference exists between using that and the IPCC-referenced simplified expressions referred to in AR6 (2021).

Its physical basis, and in particular its validity as we move on in time, has been difficult to track down. I attempted to investigate this and have summarised what I found in Appendix 3 and this has been updated in the context of IPCC AR6 2021.

Further Information:

IPCC, AR4, The Physical Science Basis, 2007
Section 2.1, FAQ 2.1 Box 1, Page136: What is Radiative Forcing
https://www.ipcc.ch/site/assets/uploads/2018/02/ar4-wg1-chapter2-1.pdf

8 - The Effect of Other Greenhouse Gasses

I mentioned earlier that carbon dioxide was not the only greenhouse gas. There are three other 'well-mixed' gasses of current significance:

1. Methane
2. Nitrous Oxide
3. A group of gasses known as CFCs

It appears that John Tyndall identified methane as a greenhouse gas in 1860. Moving forward to modern times, IPCC reports identify the sources and warming effects of all three gasses, and these will be dealt with in this section. A further greenhouse gas is ozone. This is not 'well mixed' in the atmosphere and will be dealt with in a later section.

Including Other Greenhouse Gasses

To assess the contribution of other greenhouse gasses, values were obtained for the Radiative Forcing of the gasses listed above and combined with CO2 (See Table 8.1). An early source for this data was found on the internet and agreed with IPCC values. From this table we can see that the combined forcing of these three gasses is a fairly constant proportion of CO_2 forcing. The average proportion of CO_2 to the total of all man-made well-mixed greenhouse gasses is 0.635. From this we can arrive at an estimate for the combined effect.

In the previous section, we used a reference level of 1.53% as the proportion of CO_2 absorption with respect to all greenhouse gasses (including water vapour). This did not allow for the impact of the gasses above. Because of the proportional relationship suggested above, we can simply revise our absorption to an overall figure of 2.41%. In terms of absorption, this now becomes the proportion of all well mixed greenhouse gasses, with respect only to Water Vapour. **To accommodate these gasses, the only change to our model is to change the absorption parameter to 2.41%.**

There are other ways of including the effect of the three gasses above, but these overcomplicate both the assessment and the resulting model. However, I conclude that this was not essential because the effect of the other gasses is highly correlated with CO_2. This is shown in their relationship as shown in Table 8.1. Further to this, a regression analysis shows that the effect of the other gasses is 96% correlated with CO_2. Whilst the details of regression are beyond the scope of this book, it can be appreciated that both CO_2 and other gasses are driven by the same root cause – population and industrial growth. And it is this that justifies combining their effect in a single absorption co-efficient. This aspect of correlation was covered in some detail in the previous book **(see the Section 15 in the Global Warming Handbook - on 'The Real Cause of Global Warming').**

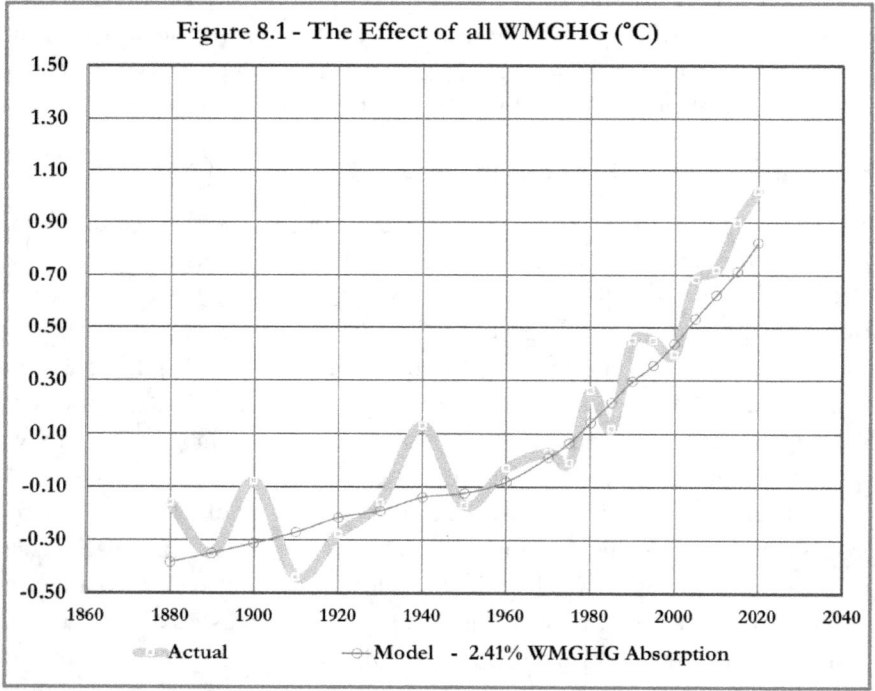

The absorption parameter is slightly larger than the value in the previous edition, containing an 18% uplift, allowing for revised 2021 AR6 methane forcing.

Looking at Figure 8.1, we can immediately see a much-improved comparison with measured temperatures.

Embedded in this are some important points:

- **CO_2 represents only 64% of the climate forcing due to man-made greenhouse gasses.**

- **When all countries work to stabilise CO_2, they will be faced with a large residual CO_2 value and must still work, both to eliminate excess CO_2 and prevent increases in the other gasses.**

Importantly, you should note that in later sections I continue to use this combined effect (including the effect of Ozone) in predictions for future years – based on the conclusion that past evidence supports a relationship between all greenhouse gases and industrial practice.

We would expect this to decrease with time. However, countries have so far fallen short of their targets. My implicit assumption is that an inability to deal with CO_2 will be matched by an inability to control the other greenhouse gasses, and the relationship may, unfortunately, hold. (to a first order, at least).

You have now reached a significant milestone. If you use the information in this and the preceding sections, you can construct a model that offers a first approximation for predicting global warming.

If you wish to move on, the sections that follow not only include significant insights into how our climate warms but also offer the first tentative links with climate change.

Table 8.1 Relationship between CO2 and other well mixed Green House Gasses

Date	CO2 Forcing W/m^2	All WMGHG Total Forcing W/m^2	Ratio of Co2 /Total	Fit of Total Forcing W/m^2
1850	0.000	0.000		
1880	0.096	0.150	**0.641**	0.151
1890	0.151	0.225	**0.670**	0.238
1900	0.210	0.306	**0.685**	0.330
1910	0.276	0.412	**0.670**	0.435
1920	0.362	0.537	**0.675**	0.571
1930	0.404	0.610	**0.662**	0.636
1940	0.486	0.726	**0.669**	0.766
1950	0.511	0.789	**0.647**	0.805
1960	0.581	0.937	**0.620**	0.914
1970	0.727	1.223	**0.594**	1.145
1980	0.937	1.643	**0.570**	1.476
1990	1.179	2.091	**0.564**	1.858
2000	1.405	2.402	**0.585**	2.213
2010	1.673	2.704	**0.619**	2.636
2020	2.013	3.090	**0.652**	3.171

Equal Time weighted assessment

1880 to 2020 Average Ratio		0.635	proportion CO2
Standard Deviation		0.039	6.1%
Proportion Forcings to CO2		1.575	proportion WMGHG
Equivalent Absorption		2.410	% including w.v.
Previous Estimate re AR5		2.371	% including w.v.

40

Further Information:

John Tyndall, pioneer in investigating infrared absorption of gasses, founder of climate science.
https://en.wikipedia.org/wiki/John_Tyndall

Human Activities and Greenhouse Gasses
IPCC, AR4, The Physical Science Basis
Section 2.1, Page 135, Frequently asked Questions 2.1
https://www.ipcc.ch/site/assets/uploads/2018/02/ar4-wg1-chapter2-1.pdf

Below is a link to the past, present and future Greenhouse gas concentrations – used as input data to the 2021 IPCC reports. The Interactive Plot presentation is excellent. You may be surprised at the slow response predicted for the future. The scenario Gcam-SSP460 looks close to the undertakings to the Paris climate conference and is middle of the road – if I have interpreted their scenarios correctly.

https://greenhousegases.science.unimelb.edu.au/#!/view

Developed by the University of Melbourne

9 – Aerosols

For some time, the topic of aerosols featured in IPCC reports. It is linked to cloud formation and scattering of sunlight by tiny particles. These extremely small particles are known as Aerosols.

Current thinking is that when moisture-laden air rises, it does not immediately generate clouds. To do this, tiny particles must act as centres of nucleation around which moisture collects to form droplets. A collection of these droplets makes up the clouds that we can all see drifting across the sky. In terms of natural sources of aerosols, it is believed that sea salt crystals levitated by wave motion are swept up in the atmosphere and are one such natural source of the tiny aerosol particles needed to create clouds.

Now when we burn fossil fuels, we also generate tiny particles that are by-products of the burning process. Similarly burning of wood fires and tropical rain forests can do the same. However, a further significant culprit is the release of sulphur compounds in the atmosphere. These also derive from burning fossil fuels as well as from petrochemical processes. If you have ever been near an oil refinery you will have recognised the smell of sulphur, particularly in past years.

Scientific expertise feeding into IPCC reports have a variety of interpretations and ultimately differing calculated effects on surface temperature.

Why is this area important? It is important because the consensus is that **aerosols act in the opposite direction to man-made greenhouse gases.** They **cool down the planet by causing scattering and reflection of sunlight.**

Including Aerosols in the Model

Up to this point, we have considered the effect of greenhouse gasses on absorption and derived the calculations needed to work out our global temperature. However, we have not considered how a spreadsheet is configured to perform the calculation and produce the graphs featured in this book. To do this, a complete calculation, starting with a measured CO_2 value, is entered - all on one line. As the line progresses: absorption; thermal Atmospheric Window Radiation; and Atmospheric Transport are calculated and then balanced against Solar Input. This results in a correction to an initial value of Ts. Each line also represents a date corresponding to the measured CO_2 value.

Whilst the value of Solar Input has been constant up until now, the value of albedo must now change on each line to allow for the aerosol effect. I have introduced the effect of aerosols by calculating an adjustment to the value of the model's albedo. This keeps our modelling in line with the physical effect. To keep things simple the albedo was changed in a linear manner from 1880 to 2020 (a rising effect) and then in a linear manner from 2020 to 2100 (An anticipated falling effect)

When revising the albedo, the overall change can be calculated using a simple formula:

ΔAlbedo = - 4 x ΔF /Sc   dimensionless

For a change in Forcing ΔF W/m²

Using Sc is the solar constant W/m²   (1366W/m²)

As an example, a forcing value of -1.2W/m² will give a change in Albedo of 0.003514.

Figure 9.1 shows how the model is configured to achieve this.

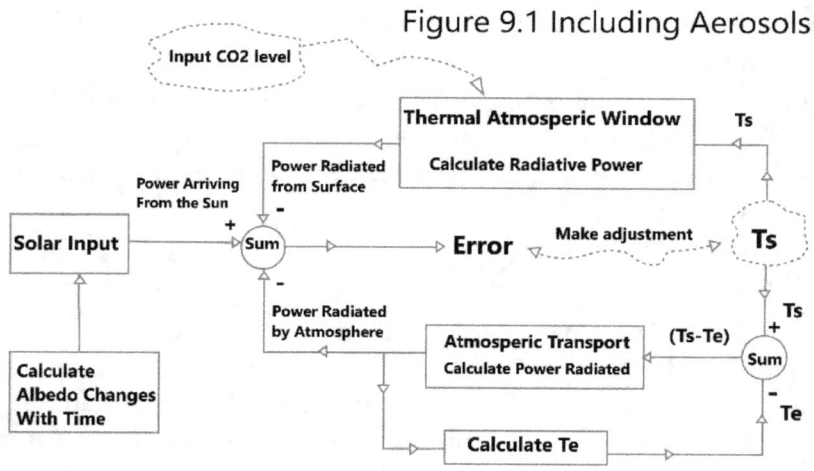

Figure 9.1 Including Aerosols

The spreadsheet allows easy adjustment of the rate of rise and fall of aerosol forcing.

I have chosen the value of -1.2W/m² for the rise in 2020 and 0.25W/m² for the for the 2100 value. This uses the satellite-derived value of -1.4W/m² reduced by 0.2 to allow for the change from 2014 to 2020. This differs slightly from the IPCC figure but lies well within its uncertainty bounds.

In Figure 9.2, you can see that the effect of including aerosols is an immediate drop in temperatures over the period from 1880 to 2020.

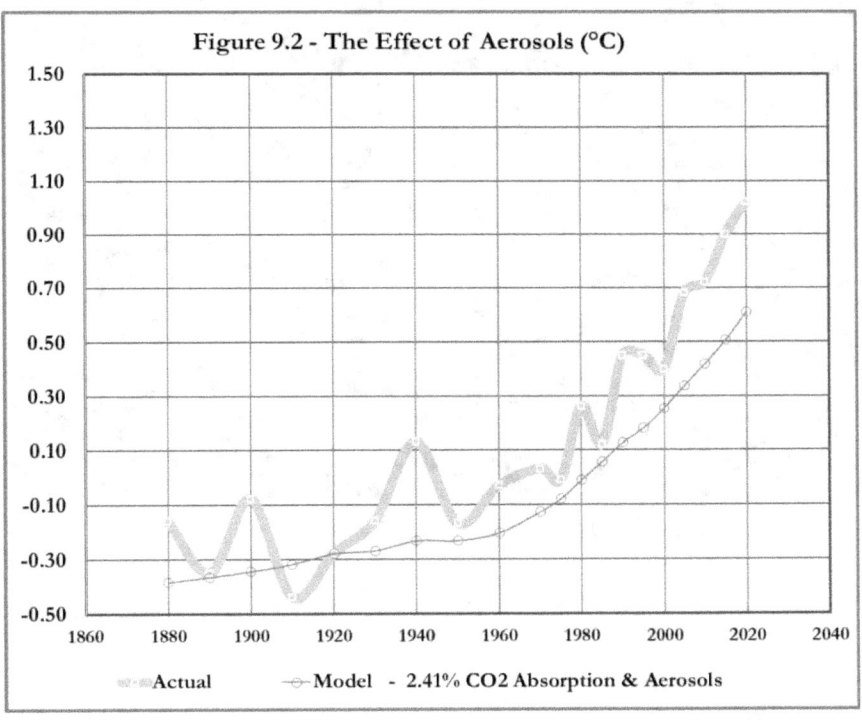

Notes:
'Clean Air' regulations are reducing emissions in many countries. If this is effective worldwide, we will see an increase in global temperature.

Figure 9.3 A Coke Production Plant in 1942

Photograph, US Farm Security Administration – Smokestacks Alfred Palmer

Further Information:

Sponsored by Bill Gates, Founder of Microsoft, is a project to cool the planet by spraying aerosols of calcium carbonate (chalk dust) at altitude. This is called Stratospheric Aerosol Injection. An article in Nature covers this in more detail.
https://www.nature.com/articles/d41586-018-07533-4

10 - Temperature Feedback

We have seen in the previous sections that when we use physically based absorption data together with an estimate of the aerosol effect, that there is a shortfall in our model's estimate of temperature rise. This means that something else must be driving measured temperatures upwards. The likely candidate is termed Temperature Feedback. This is an important area because, it can be responsible for a large additional increase in temperature It is also important because it can yield an estimate for the amount of additional water vapour in the atmosphere – at which point we will have in our possession two of the most important parameters in predicting changes in climate: Temperature and Atmospheric Water Vapour.

Temperature Feedback means that a physical process amplifies (or reduces) the temperature change caused by greenhouse gasses. Within the atmosphere, several potential feedback processes have been identified. The most significant is the interaction between increasing temperature and 'atmospheric water vapour'. Climate scientists refer to this as Water Vapour Feedback. At this point it is worth pointing out that several feedback processes can feed on each other, growing larger and larger. In many ways, it can appear a daunting prospect because it raises the spectre of a run-away temperature rise.

Water Vapour Feedback

Water Vapour is the most overwhelming Greenhouse gas. Air is capable of holding water vapour, and in our day-to-day lives, we feel this directly as humidity. The warmer the temperature, the more water vapour can be held in the atmosphere. Should the quantity of water vapour increase beyond its saturation limit, then the atmosphere gives up the excess amount of water – it probably creates clouds, rains, or snows, or produces condensing fog. But the bottom line is that this humidity limit cannot be exceeded – the atmosphere can hold no more water vapour than its saturation limit.

However, when CO_2 causes our atmosphere to warm, the atmosphere can hold more water before saturating. In the right conditions, it will suck up more water until some form of equilibrium exists. This is called evaporation. This process will occur anywhere on Earth where water is available, but the tropical ocean surfaces clearly win 'hands down'.

Heat must be supplied when water changes state from liquid to vapour. This is called Latent Heat, and a large amount of heat is required to achieve this transition. We all have personal experience of this. When we are hot, we sweat. This moisture evaporates from our skin, and Latent Heat is drawn from us. As a result - we cool down. Also, we know from our washing line that if the atmosphere is humid, evaporation works poorly, and our clothes dry slowly. If there is a breeze, evaporation speeds up, and we dry our clothes quickly.

To enhance global warming, we first need the processes discussed in previous sections to simply warm the atmosphere. In its warmed state it can absorb more water vapour. However, we must also identify a new source of heat to evaporate the water.

You might think that stored heat is simply extracted from the ocean – but the ocean and its currents are part of the Atmospheric Transfer heat transfer process – At first sight this results in no net heat gain and no temperature amplification.

However, the analysis I suggest is one where sunlight supplies heat to the ocean, which in turn heats a boundary layer of air. The top of this boundary layer is where infrared is radiated directly into space via the thermal atmospheric window, as explained in section 4

Evaporation also takes place in this boundary layer. Inevitably, the surface of this boundary layer will be cooler than it would otherwise have been (latent heat is extracted). This cooling must result in less radiated energy via the thermal atmospheric window. If less heat escapes via the thermal atmospheric window, then our overall atmospheric temperature must increase, and more heat enters our atmosphere.

The process becomes one of self-amplification or positive feedback. It is also one where a balance is struck, between chilling of the boundary layer on the one hand and evaporation on the other. At first sight it is not easy to see how to evaluate the equilibrium condition.

At one level, we could just look at IPCC reports for the value of feedback, but this provides little insight into the basic physics. It means that our personal ability to predict ahead is limited by information that we do not possess because it will lie embedded in a complex computer simulation, and a mass of three-dimensional partial differential equations. These simulations yield a solution across the entire troposphere and points to a warming of the upper troposphere as a key part of the water vapour feedback amplification process.

I am going to suggest a pragmatic approach. An approach, with the aim of keeping things as simple as possible but still physically based. Because this can appear daunting, I have split up how this is presented.

In this main text is a simple presentation of how to change the model to allow for Water Vapour Feedback using two coefficients. In two appendices, I will provide the analysis that supports this. You can choose the depth you wish to engage in this. However, in the appendices, you will find an analysis approach that I have not seen before. It uses only basic algebra, some measured data, one line of simple calculus, and a bit of thought (quite a bit). However, it offers something else. It enables you to take a first step towards evaluating Climate Change. It does this by providing a direct evaluation of the additional amount of water vapour entering the atmosphere. With this evaluation we possess the means to model the two key parameters that influence our climate: Global temperature rise and the global increase in water vapour.

To enable this approach, I have split the modelling of Water Vapour Feedback into two parts:
- Evaporative feedback – which estimates the amount of evaporation at an ocean surface boundary layer.
- The resultant increase in atmospheric heat capture - due to increased water vapour.

Evaporative Feedback

First, we need to estimate the increase in atmospheric water vapour. If we assume a thin evaporative boundary layer (which Appendix 5 estimates at 240 microns), we can also assume that it is saturated. So, our starting point is to calculate the change in Water Vapour Saturation Density when we increase global temperature (Saturation Density is the mass of water held in a cubic meter of air.) We can calculate this easily using a polynomial. (see Appendix 5). We can do this for each line of our spreadsheet (each value of greenhouse gas heating), and as we progress from one line to the next, we then know both the change in saturation density (**Δm**) and its former absolute value (**m**). This represents the maximum potential change in vapour density.

For a change in vapour mass, we require a supply of latent heat. As explained above, this is a delicate balance between cooling, radiative loss via the thermal atmospheric window and evaporation. Appendix 5 explains how to derive this balance, with the outcome that for a change in temperature, with a corresponding change in vapour density:

Feedback = +8.99 W/m^2 for changes in (Δm/m)

Where **Δm** is a change in vapour density compared to its original value **m**.

Figure 10.1 provides an extract from the spreadsheet showing this calculation. Successive lines of the calculation are for successive values of CO_2. With the (not shown) preceding part of each line calculating the value of Ts (Global Surface Air Temperature) for successive CO_2 values. In the extract, you can see a calculation of the vapour density on each line, and the difference in vapour density between lines is the global warming-induced change in saturated vapour density. When this is multiplied by the factor above, we have a value of the power to be fed back (I actually deduct it from the thermal atmospheric window heat loss). This, in turn, causes more global warming as the spreadsheet iterates to a solution. As the spreadsheet progresses from 1880, to say, 2020, the feedback is cumulative after each change in CO_2 and corresponding Global Average Air Temperature, Ts.

At this point I should explain a little further. This approach does not assume that the atmosphere remains saturated above this small boundary layer. Away from the boundary, it will assume a non-saturated value – as mixing with the atmosphere takes place. However, that does not need to be explicitly evaluated using this approach. Instead, it uses measured data from ocean buoys to establish the boundary conditions that replace the need for complex simulation.

Further, the heat transfer conditions are evaluated without the need for knowledge of a specific heat transfer process.

Water Vapour Feedback

Saturation density and rate of change with T

Water Vapour Feedback

Work out water vapour mass change

Revised Ts K	Temp. °C	Change of saturation density gm/m³/K	Mass at saturation density gm/m3	delta T °C	Delta m gm/m³	Delta M/M	For Info (DM/m)/DT K⁻¹	change in power Watts/m²
286.83	14.18	0.744	12.138					
286.86	14.21	0.745	12.159	0.029	0.0213	0.0018	0.0613	0.016
286.89	14.24	0.746	12.183	0.03	0.0237	0.0020	0.0613	0.018
286.93	14.28	0.748	12.212	0.04	0.0294	0.0024	0.0613	0.022
286.99	14.34	0.750	12.255	0.06	0.0429	0.0035	0.0614	0.032
287.00	14.35	0.751	12.268	0.02	0.0125	0.0010	0.0613	0.009
287.06	14.41	0.754	12.308	0.05	0.0402	0.0033	0.0613	0.029
287.06	14.41	0.754	12.309	0.00	0.0012	0.0001	0.0612	0.001
287.10	14.45	0.755	12.341	0.04	0.0318	0.0026	0.0613	0.023
287.21	14.56	0.760	12.426	0.11	0.0851	0.0069	0.0614	0.062
287.28	14.63	0.763	12.480	0.07	0.0537	0.0043	0.0613	0.039
287.39	14.74	0.768	12.557	0.10	0.0776	0.0062	0.0614	0.056
287.48	14.83	0.772	12.633	0.10	0.0757	0.0060	0.0613	0.054
287.59	14.94	0.777	12.713	0.10	0.0800	0.0063	0.0613	0.057
287.67	15.02	0.780	12.775	0.08	0.0617	0.0049	0.0612	0.044
287.77	15.12	0.785	12.859	0.11	0.0842	0.0066	0.0613	0.059
287.90	15.25	0.791	12.958	0.13	0.0986	0.0077	0.0613	0.069
288.02	15.37	0.796	13.053	0.12	0.0951	0.0073	0.0612	0.066
288.15	15.50	0.802	13.154	0.13	0.1012	0.0078	0.0612	0.070
288.30	15.65	0.809	13.276	0.15	0.1219	0.0093	0.0612	0.083
290.67	18.02	0.922	15.326	2.37	2.0505	0.1545	0.0651	1.388
291.06	18.41	0.942	15.688	0.39	0.3620	0.0236	0.0608	0.212
291.99	19.34	0.990	16.595	0.94	0.9063	0.0578	0.0616	0.519

Figure 10.1

This approach was adopted to produce the graphs in the Global Warming Handbook.

Figure 10.2 shows the revised energy balance diagram. The box containing Water Vapour Feedback includes the processing outlined above as well as the next two processes in the text that follows.

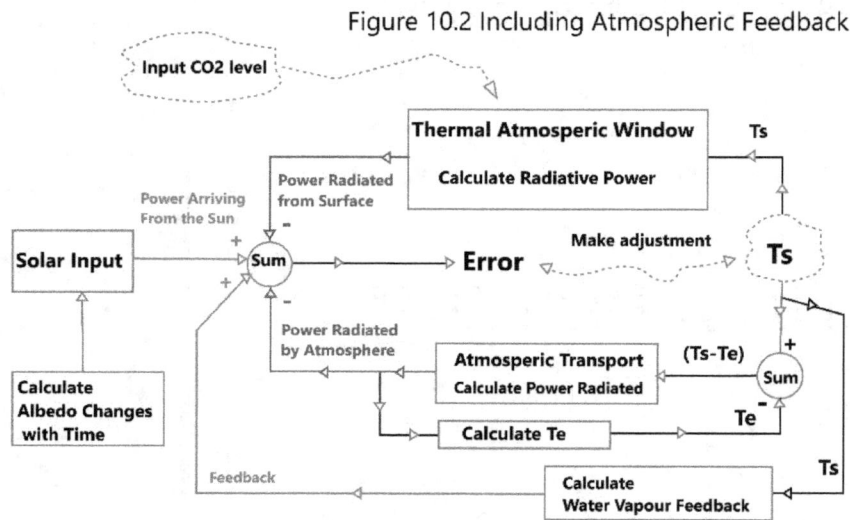

Figure 10.2 Including Atmospheric Feedback

Increased Atmospheric Capture of Sunlight

The evaporative feedback above is only part of the "water vapour feedback" story. Once we have an increase in evaporation, we have an increase in atmospheric water vapour.

There is no change in terms of the earth's capture of heat by this process. Incoming sunlight is either captured by the atmosphere or the earth's surface (land and sea). The total heat capture remains the same. However, it does reduce the proportion of heat directly absorbed by the Earth's surface. This, in turn, reduces the amount of thermal radiation from the Earth's surface. The outcome is that less cooling takes place. Appendix 6 covers this in more detail and arrives at an estimate for this feedback of +0.3 W/m² /°C.

You need to be careful at this point. The evaporative feedback is calculated based on a change in evaporative mass. The water vapour heat capture feedback is evaluated on changes in temperature.

Increased Absorption of Window Radiation

This time it is an increase in the absorption of thermal radiation that needs to be considered.

Appendix 6 investigates this and concludes that the effect would be negligible, as water vapour absorption in this window region appears saturated, and CO_2 levels are unaffected.

Atmospheric Feedback Roundup

In the preceding sections, we have looked at Water Vapour Feedback and identified three components:

- Evaporative Feedback
- Increased Atmospheric Capture of Sunlight
- Increased Thermal Radiation Capture

Where relevant, we have arrived at a numerical value for each of these, using measured data as a basis. It is worth taking the trouble to do this because water vapour feedback is such a major contributor to global warming. Sometimes, it is attacked as a "convenient fiddle" (generated to explain global warming). Something lost in the depths of some giant computer, guarded by the high priests of climate change. Yet we have independently managed to produce an overall approximation – and it is an approximation. But we will draw some encouraging comparisons with IPCC figures a little later, where the Water Vapour feedback resulting from using this approximation agrees with the AR6 'mean' figure almost exactly.

However, Water Vapour Feedback is only one part of overall Atmospheric Feedback. Two other significant processes are involved:

- Temperature Lapse Feedback $-0.5 \ W/m^2/°C$
- Cloud Feedback $+0.42 \ W/m^2/°C$

I am not going to divert to analyse these, but suggest we simply use the IPCC AR6 values. Intriguingly, when we add these together, they almost cancel out (Using AR5 figures, they did cancel out). Leaving evaporative feedback and Water Vapour absorption as the atmospheric feedback term. One of the reasons to include these values, as well as increased atmospheric sunlight capture, is to allow them to be changed as information quality improves.

Figure 10.3 shows the effect on the model of Water Vapour Feedback: an improved agreement with measured temperatures.

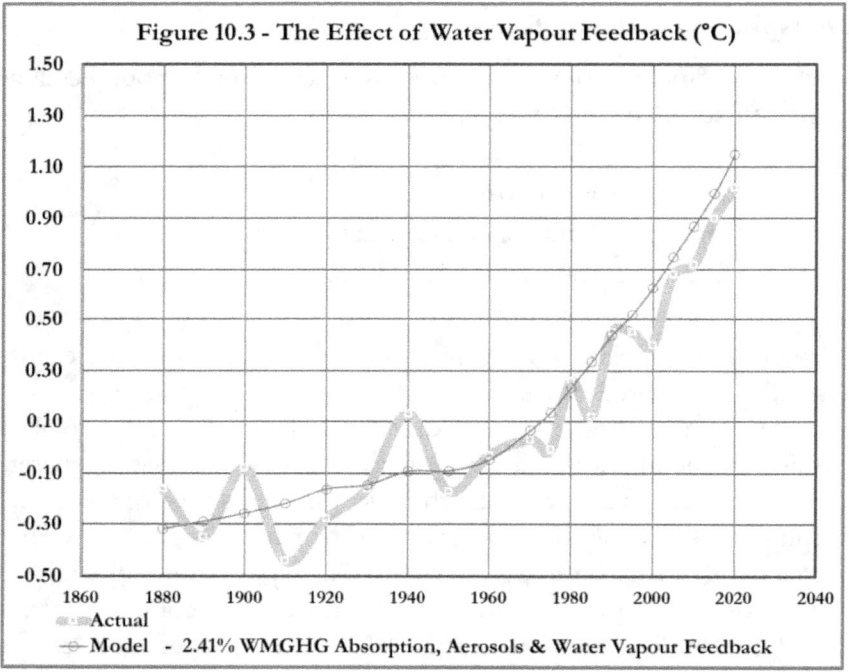

Increased Atmospheric Water Vapour

Finally, using this model of Water Vapour Feedback enables us to evaluate the increase in atmospheric water vapour due to Global Warming. We can do this because we know the saturated water vapour density at the ocean surface in 1880 and at any date where we model. As this represents the water vapour entering the atmosphere, a simple ratio tells us how it changes from one date to another. Figure 10.4 is a plot of this ratio from pre-industrial times to the present. This indicates an overall increase in atmospheric water vapour of more than 9% by 2020.

I found it difficult to obtain a published report of overall water vapour content to yield a comparison. The IPCC report I included under further information) yielded a figure for the tropics of between 1.5% per decade to 3.7% per decade for the period between 1965 to 2000. This suggests an increase in this period of between 5.25% to 12.95%. This range shows the difficulties in measuring this parameter across a range of altitudes. But **it does give outline confidence in the model's assessment which agrees almost exactly with the <u>mean</u> of the IPCC report over the dates in question.**

Water vapour content is important because this, together with temperature, determines our climate. In combination, they determine rainfall trends and aridity. Complex climate modelling does give us insights into future trends. This suggests that wet areas (the tropics) get wetter, and desert areas extend. If you wish to look into this further, you could also look at the Koppen system, which informs us of the current climate distribution.

Looking again at Figure 10.4, you can see that it predicts that by 2100 (Greenhouse gas levels for 2100 are covered in section 13), the atmospheric water content may rise by 29% (26.4% in the later final model) from pre-industrial conditions. It is one thing to look at what seems to be small temperature changes and dismiss them. It is another to look at an almost threefold change in atmospheric water content from 2020 to 2100 without trepidation. Already, in 2021, increases in storm intensity and flooding in some areas, together with extreme heat in others, are being suggested as linked to global warming.

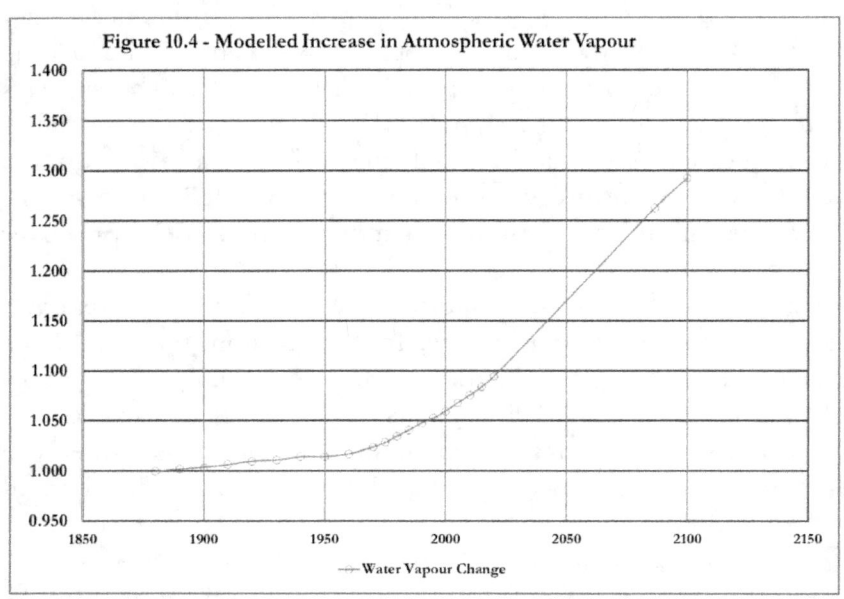

Figure 10.4 - Modelled Increase in Atmospheric Water Vapour

Further Information:
1/ Water Vapour Feedback and Global Warming (2000)
I M Held and B J Soden (look at the introductory historical section)
https://pdfs.semanticscholar.org/4d51/4320c77e2e5fd5ce231c1adbe287649b6356.pdf?_ga=2.127967948.1321871591.1568995130-28063494.1568995130

Notes:
- *Increased water vapour translates to heat carried in the atmosphere, not just water content.*

- *Attributing climatic events to Global Warming not easy and avoiding speculation can be difficult. Below is a link to an attribution study which concludes that the European* **'July 2019 heatwave was about ten times more likely because of climate change'**
 https://www.metoffice.gov.uk/research/climate/understanding-climate/attributing-extreme-weather-to-climate-change

11 – Ozone and Other Forcings

This section deals with a number of omissions from the previous edition of this book, as well as Surface Albedo Feedback:
- A group of Other Forcings
- Ozone

I will deal with them here as well as Albedo Feedback.

Other Forcings

These contribute a small amount to global warming but are highlighted in the summary table of Forcings in AR6. I have included them for completeness.

	W/m^2
Land Use This deals with the changes in surface reflective properties due to human activities. For example, replacing forests with more reflective land surfaces.	-0.20
Contrails The creation of aviation-induced cirrus	0.06
Surface Albedo E.g., black carbon deposits on snow and ice	0.08
Total	**-0.06**

These effects are caused directly by industrialisation and are not responses to climate change (also, they are not feedback terms). As a minor contribution, they can reasonably be accommodated as responses to CO2, which is correlated with industrialisation. For convenience, they have been introduced as a very small change in the worth of greenhouse gases, as discussed in Section 8.

Surface Albedo Feedback

This might well seem confusing. We have just been introduced to Radiative Forcings that sound very similar under 'Other Forcings' However Albedo Feedback is a direct response to climate change. As temperatures respond to climate change this term caters for the feedback response. For example, as snow cover is affected, it will reduce reflections of incoming sunlight and increase absorption of heat. This will raise global temperature, causing more loss of snow cover.

As far as practical, my aim is to model changes in terms of the physical parameters they affect. In this case, the albedo term of the model needs to be adjusted.

To achieve this, a feedback loop is set up so that the global temperature response is reflected in a change in forcing, which the model translates to changes in albedo and so on until an equilibrium is established.

Albedo is already adjusted for aerosols by generating a time ramp of albedo changes from, say, 1880 to 2020. Feeding a temperature-adjusted forcing into this system produces the necessary feedback loop. To achieve this the IPCC AR6 albedo Feedback value of 0.35W/m²/K is modified by the modelled temperature at 2020.

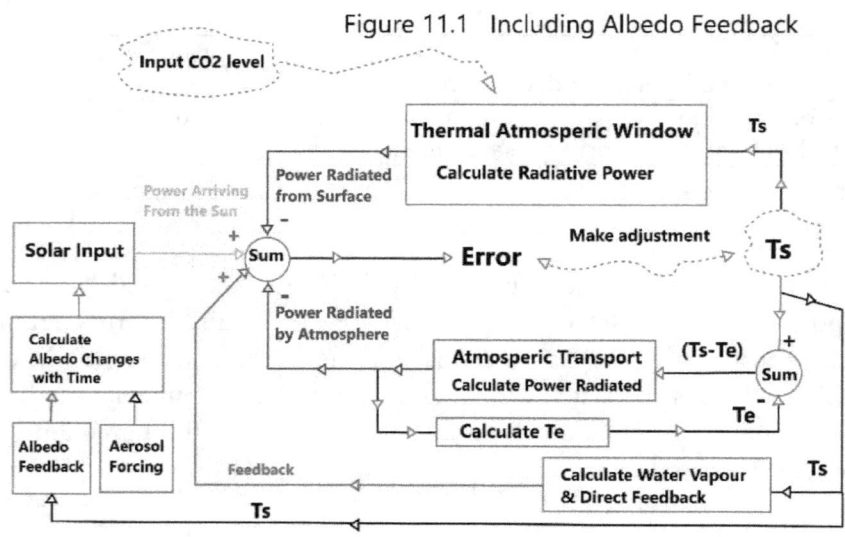

Figure 11.1   Including Albedo Feedback

Ozone

Ozone is present in our atmosphere in two areas: the Stratosphere and the Troposphere. In the stratosphere, it is best known for its depletion:

by man-made chemicals. The so-called 'Hole' in the atmosphere causes a harmful increase in ultraviolet light. In the troposphere, it is an important greenhouse gas. It is not man-made but is caused by the interaction between man-made chemicals and sunlight. It is not well mixed in the atmosphere and is transitory in nature. This makes it difficult to assess by observations. All assessments of its impact on global warming (past and present) are arrived at by calculation and contain significant uncertainty. AR6 assesses its Radiative Forcing as $0.47W/m^2$ from 1750 to 2018. To align this with our model, which runs from 1880, I have simply deducted $0.03W/m^2$. The question is how to apply it. Noting that Methane and Nitrogen Oxides (from car emissions) are precursor chemicals, it would seem pragmatic to include it in the WMGHG forcing calculation. This approach is supported by its creation of 'products of industrialisation', which are correlated with CO_2.

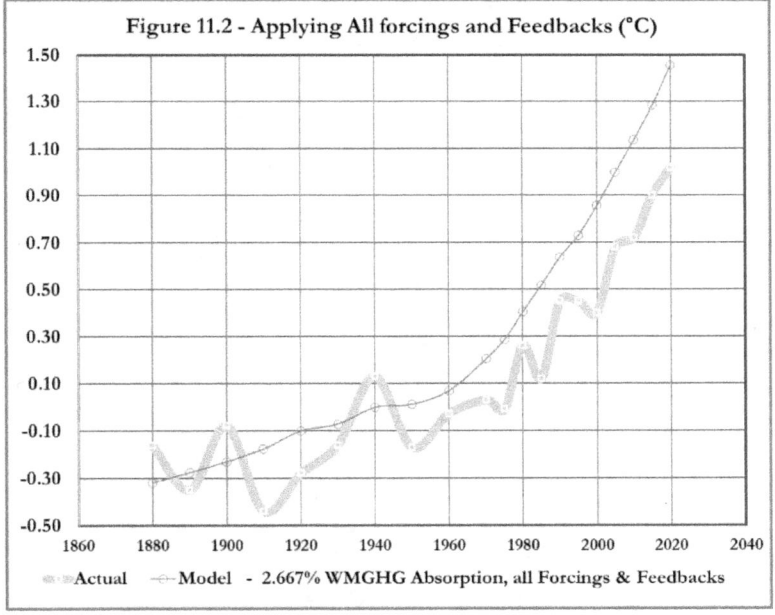

Figure 11.2 - Applying All forcings and Feedbacks (°C)

The outcome is a change in the percentage worth of WMGHG to 2.665% of water vapour and Figure 11.2.

Having applied all the currently reported Radiative Forcings and Feedbacks, we have arrived at a model that suggests our current Global temperature should be 1.5 °C (much higher than we measure). We must have some hidden protector – because our actual measured temperature is much lower.

12 – The Oceans and Energy Balance

The preceding sections assume that a heat balance is reached between the heat coming into the planet and that leaving. In practice, it will take time for the planet to adjust to a change in Greenhouse gases. It cannot do this instantaneously. Scientists refer to this aspect as Dynamic behaviour.

The dynamics of Climate Change are phenomenally complex. This is reflected in the fact that the IPCC report (AR5) used the results of 30 models, and AR6 used more than 50 to evaluate climate change sensitivities. These models differed in complexity and approach and produced a range of results. They are clearly thought necessary to cover the spread in thinking on the subject.

I will draw out a few basic points that affect the timing of Global Warming to establish the role of the oceans.

First a few essential points about heat storage. If you heat something up, its temperature will change at a rate depending on something called its thermal inertia. We see this in everyday life. A concrete slab heats up and cools down at a very much slower rate than, say, a cup of coffee. Put simply, the slab has a greater mass – in fact, a greater thermal mass. Similarly, the world's oceans are massive. They account for over 70% of the earth's surface and descend to unimaginable depths: they hold a lot of water.

Whilst sunlight affects the surface of the oceans, Oceanographers determine that this heat is distributed within the top 500m by waves and wind. In this section, I use a nominal depth of 200m to describe this region.

You might think that the land is solid and must store a lot of heat. However, the equivalent depth of land is suggested by some sources to be about 1.7m. Taking other factors into account, a quick estimate suggests that the thermal inertia of the land is **hundreds of times smaller** than the top 200m of ocean. This is partly because downward conduction through land is poor.

Surprisingly, the atmosphere, all that light fluffy stuff, has a greater thermal mass than the land and can be estimated at about **75 times less** than the oceans. The details of these calculations can be found in the model spreadsheet.

To add to this, the oceans don't reflect as much sunlight as the land (about 6% reflection compared to about twice this for land surfaces).

Putting this all together, the oceans capture the bulk of the planet's energy from sunlight and are great at storing it. This is reflected in the National Oceanic and Space Administration's (NOAA's) estimate of mean global mean ocean temperatures from 1800 to 2000. They estimate a contemporary average air temperature above the ocean of 16.1 °C compared with the average air temperature above land of 8.5 °C.

The oceans and their interaction with the atmosphere primarily determine the rate at which our world heats up.

A simple dynamic model of the mixed ocean layer yields a time constant of about 4.6 years – a time constant is the time to react to 63% of equilibrium for say a step change. Whilst the land will have a time constant of a month or so. (Look at the notes at the end of this section)

Most importantly, the ocean depths extend to a depth of 3.7 kilometres (not metres) on average. These depths will heat up very slowly, and, in the meantime, there will be a barely perceptible change in temperature. To understand the effect of this, we need to know how much heat penetrates to these depths.

As the ocean depths absorb heat with very little to show for it, this can only be revealed in an energy balance calculation. In other words, a certain amount of energy comes into the planet, and a certain amount goes out. If energy is stored without noticeable temperature change, some will seem to disappear. Scientists feeding into AR6 have evaluated a contemporary Global Energy Imbalance of $0.7 W/m^2$. A very small amount, and if we use this to calculate the deep ocean's time constant, we can obtain a value of 6500 years.

Using this together with a further estimate of a preindustrial imbalance of $0.2 Wm^{-2}$, we can formulate a simple heat transfer relationship.

Imbalance flux (W/m^2) = Ki (Ts-Tref)

We can solve this for the preindustrial and current imbalance figures to give a value of T_{ref} and Ki.

If W_{f1} is the preindustrial imbalance and W_{f2} is the contemporary imbalance, then:

$W_{f1} = K_i(T_{s1} - T_{ref})$ W/m^2

$W_{f2} = K_i(T_{s2} - T_{ref})$ W/m^2

Therefore, as K_i is common to both and assumed constant – we can equate

$W_{f2} / (T_{s2} - T_{ref}) = W_{f1} / (T_{s1} - T_{ref})$

And rearranging we have:

$T_{ref} = (T_{s1}[W_{f2}/W_{f1}] - T_{s2}) / ([W_{f2}/W_{f1}] - 1)$

This yields a value of 13.494 °C. Taking this further we can look to the future in 2100 and using an estimate for the Global Surface Air Temperature (Ts in this book)) we can evaluate the imbalance at this future point. Clearly this will depend on our assessment of Ts in 2100.

(1) This eanbles the heat transfer Co-efficients, Tref and K to be evaluated
Adjust Tref such that K_{th} is the same for 1880 and 2018
Alternatively Insert Calculated Tref

Calculated Tref:		13.4940 Deg C		Tref (°C)		Ki (Wm^{-2}K^{-1})
				13.494	°C	0.48077
				286.644	k	
date	imbalance		GSAT	T$_{Ref}$	ΔT	
	W/m2		Deg C	Deg C	Deg C	
			smoothed			
1880	0.200	Wm^{-2}K^{-1}	13.91	13.494	0.416	0.48077
2018	0.700	Wm^{-2}K^{-1}	14.95	13.494	1.456	0.48077
calculate future conditions						
2100	1.945	W/m2	17.54	13.494	4.046	0.48077
	Result		Trial			

Suggesting that   13.5   Deg C   represents a point in time where,
prior to 1850, the imbalance was zero   (GSAT is Global Surface Air Temperature)

Figure 12.1

Figure 12.1 shows an extract from the model spreadsheet. From this, we can infer that Tref represents a surface air temperature – when, sometime in the past, the ocean/ air temperatures were in equilibrium. We can also see that the forcing in W/m² varies with surface air temperature, and the coefficient Ki has units of W/m²K⁻¹: the units of a feedback term.

The above relationship is conditional on the ocean abyssal depths slowly changing in temperature - such that Tref remains essentially constant. This condition is satisfied by the time constants referred to above. You should note that this reveals an increasing "forcing" as global temperature rises. This is a valuable and increasing reduction of global warming impact. It could be worth almost 2 Wm⁻² by 2100, with the temperature increase noted in Figure 12.1.

Our model up to this point shows a large discrepancy between measured and modelled temperature. Our model's prediction was much hotter than actually measured. However, it did not take into account the heat lost to the oceans. This is easily included by using a specific feedback term based on ki and Tref

When this is done, this result is an agreement between model and measurements and is shown in Figure 12.2.

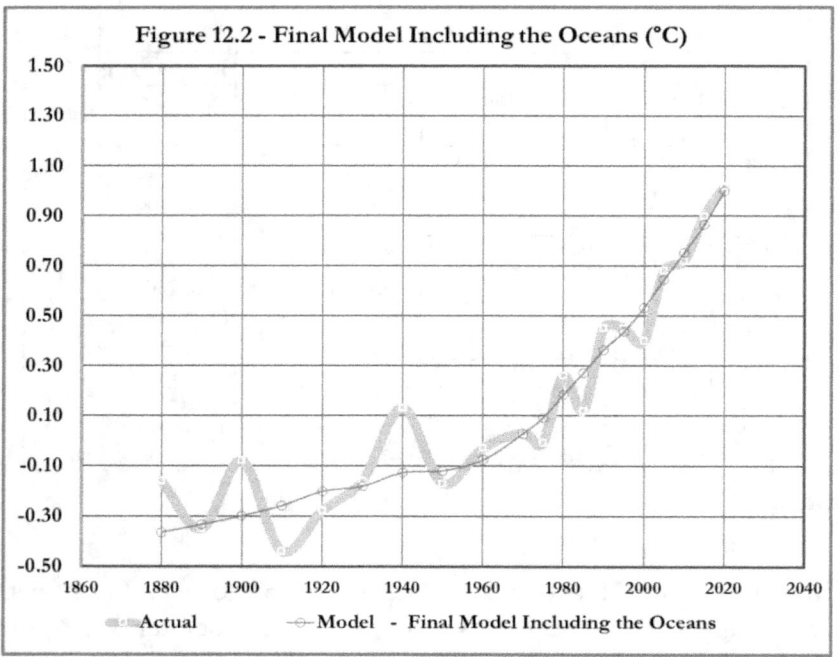

The result is fortuitously good – given the wide range of parameters (often with high uncertainty). But it does, in general terms, reveal the protective thermal effect of the oceans.

Notes:

- *I have referred to a simple dynamic model in assessing various time constants. This is outside the scope of this book. However, several relevant dynamic models are explained in a further book linked to this series:*
    - **Simulation Made Easy, By Jim Wiles available from Amazon.**

- The oceans are a fascinating subject in their own right, and the treatment above, in seeking simplicity, barely touches their complexity. For example:
    - Their temperature gradients vary with latitude
    - Their deepest points are a few degrees above freezing
    - Their deepest points are subject to geothermal heating, the significance of which is now recognised
    - The wind-driven gulf stream flow cools and sinks to the depths to re-emerge a thousand years later in the Southern Ocean.
    - Truly a great research area to tackle, with such significance to planetary warming and our future.

(Continued overleaf)

Satellite Measurements play a key role in establishing the Earth's Energy balance. The joint Polar Satellite System (JPSS) is a collaborative program between NOAA and NASA. Measuring atmospheric, terrestrial, and oceanic conditions.

13 – Model Prediction Current and Future

Comparison with Measurements

One of the key tests of any model is how well it matches actual temperature measurements from preindustrial to current times. This is most clearly shown in Figure 13.1. Over the period from 1880 to 2019 the average deviation over this period is 0.004 °C. However, for predicted temperature deviations, which vary about some average value, it is better to use something called the Standard Deviation as a measure of agreement. For this model, the standard deviation is 0.11 °C. Further the model accounts for 93% of the variability in the measured data (admittedly only 19 data points are used).

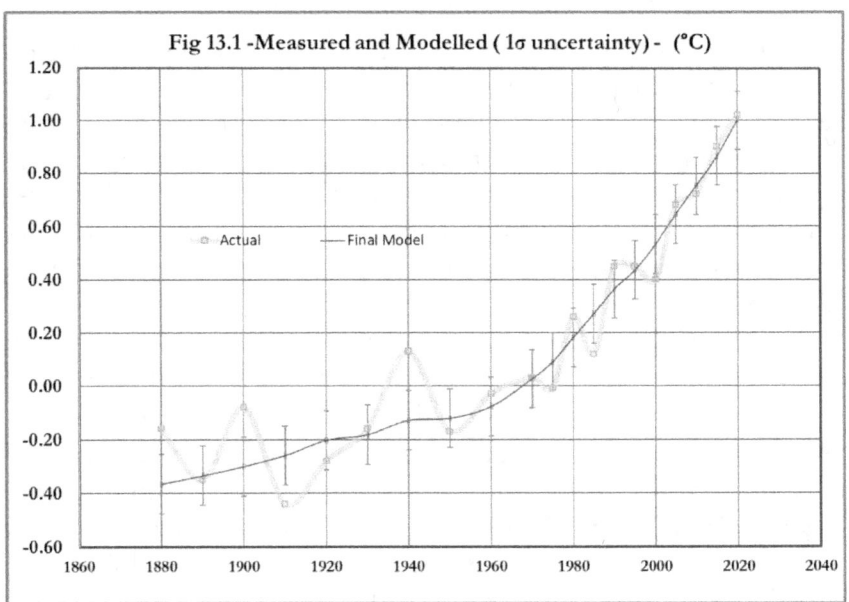

No corrections or tweaks have been applied to this model to achieve the outcome. I have referred to the result above as fortuitously good, but just to recap on how we got here.:
- The outcome is based on a physically based model.
- It does make use of data that is contained in IPCC references.

- However, it also makes use of measured data wherever possible and uses limited judgement about the value of parameters used.
- It independently derives an expression for Water Vapour Feedback that can be used forward in time and implements the effect of this.
- It independently arrives at an expression for the effect of the oceans that can be used forward in time.

For what started as an illustrative model, the. approach of using a physically and simply based model has proved to be successful in predicting contemporary temperature changes.

Future Prediction

Moving on to future predictions. In 2015 a climate conference was held in Paris where each nation gave undertakings on how they would reduce greenhouse gas emissions. If these were adhered to strictly, then this would result in an increase in atmospheric CO_2 - reaching 670 ppm by the year 2100 (the National Plans prediction). This, in turn, would result in an estimated increase in Global Mean temperature of 3.5 °C —from **pre-industrial times**. This "National Plans" response was viewed as inadequate, and goals of 2°C and then 1.5°C were put forward.

However, delivery against their objectives reported at the end of 2019 indicated that unless drastic changes were brought about, we were inexorably heading for the National Plans target of 3.5°C by 2100.

A sensible test of our model is to see how it compares with IPCC estimations, and it seems reasonable to use the National Plans prediction as a comparison. Figure 13.2 depicts the National Plans prediction as an extension to the measured temperature changes (the broad line). Note the pre-industrial baseline adds around 0.2°C to the anomaly.

Of course, the assumptions on greenhouse gas levels, other than CO_2, may be different in this comparison. The assumption used in our model is that the Nations' response to other greenhouse gasses will be similar to their ability and motivation to control CO_2 releases.

Both our model and the 'National Plans' estimation use the same value of 670ppm CO_2. IPCC scientists have done the hard work in coming up with the CO_2 value, which our model processes into temperature change. Our model gives a prediction of 3.7°C for a preindustrial baseline (within 7% of the 3.5°C IPCC value) and is encouraging. It does illustrate that the model used in this book is representative of conditions extensively modelled by others.

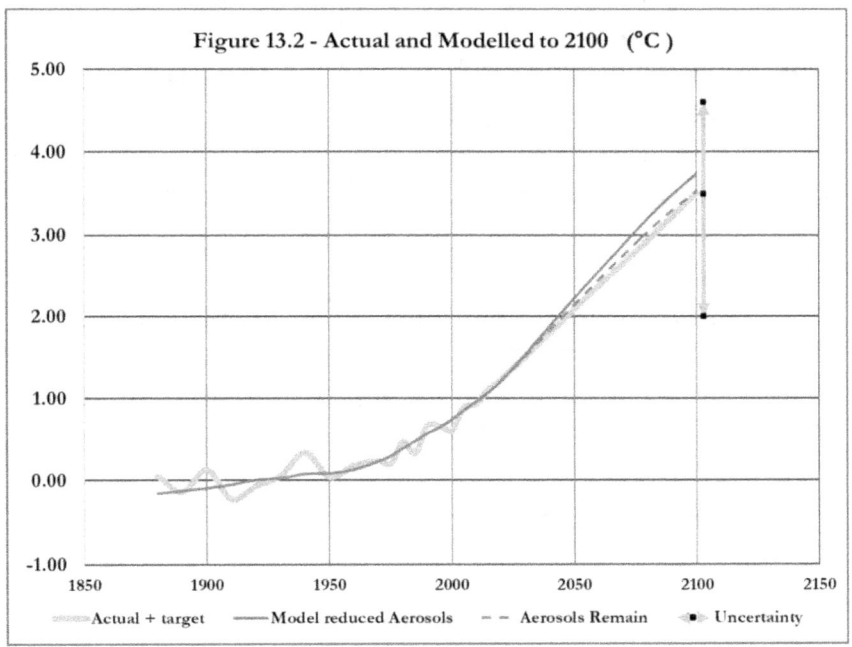

At this point I should stress again that at no stage was the model constructed to reproduce IPCC findings. Further, the analysis that underpins the model has been derived independently – (I am sure that others must have been down similar routes, but I have neither located nor used their work). The model does use IPCC information on Radiative Forcing - derived from gas concentrations, largely measured, and absorption based on commonly available spectrographic codes. Some critics of global warming often criticise Atmospheric Feedback as 'some sort of fiddle to get a big global temperature rise'. In some ways, this is understandable, as climate scientists rely on supercomputer models that must be opaque to most of us. However, in this book we have independently arrived at a unique approximation – and it gives a comparable outcome to IPCC models.

It is also intriguing to look back again at the review of Aerosols in Section 9 and see what happens as man-made aerosols are reduced. We can also hold the Albedo at the 2020 level if we choose, and this gives us an estimate of the warming due to aerosol reduction. This is also shown in Figure 13.2. The dashed line shows that predicted temperatures would have been 0.2°C cooler if industrial aerosols had still been present or stratospheric injection deemed beneficial and implemented. Bear in mind that the aerosol effect is now assessed in AR6 at a larger value than AR5 (as suggested in the first edition of this book.) In any event, the aerosol effect is still subject to large uncertainty, although AR6 has now evaluated limiting values, within which our assessment lies.

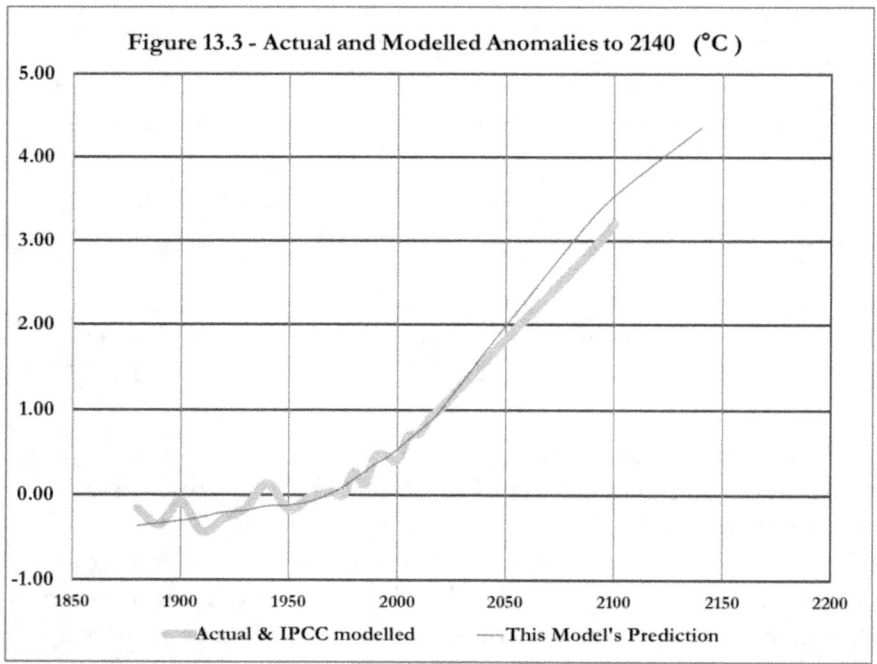

Note: This graph reverts once more to the mid-20th Century baseline

Intriguingly, **our model does not show a runaway situation** because of temperature feedback. In fact, this did not happen at higher CO_2 values. The reason for this is best illustrated in Figure 13.3, where a token value of 799ppm CO_2 is fed into the model. Here, you can see the modelled line starts to bend downwards. Presumably reflecting negative feedback (from atmospheric and thermal window radiation) - Potentially with increasing CO_2 saturation effect and Ocean heat uptake.

Notes:

- *Looking ahead to 2140 with our model, there is no sign of a runaway temperature rise. However, it should be noted that IPCC AR6 predicts an increase in climate sensitivity with CO_2 level, although not stated as a runaway.*

- *This section has used data relating to the Paris Climate Conference. This serves as a basic crosscheck of the model presented here. Since then, scientists have doubted if even the 3.5 °C projection for 2100 could be achieved.*

- *The outcome of the 2021 Glasgow COP26 conference has had a mixed reception, with positive undertakings undermined by poor perceived overall impact. However, Carbon Brief has suggested an outcome with temperatures reaching 2.6 to 2.7 °C (with an uncertainty stretching up to 3.6 °C). They also suggest that Methane and Coal pledges may only result in a 0.05 °C reduction in temperatures by 2100. As yet I have been unable to find atmospheric CO_2 projections to match the COP26 outcome. When they arise, you may be intrigued to enter these into our model.*

14 – Uncertainties

The model described here was established as an illustrative model - something that could credibly represent the major processes that influence global warming. To do this it had to simulate those processes in simple physical forms. The objective was to engender an understanding of those processes to a wider audience. In the first instance this was presented via the Global Warming Handbook, using a straightforward and limited level of technical explanation.

In order to be a credible illustration, it is reasonable to ask, 'how accurate is the model?'

There are several aspects to this:
- The dependency of the model on other work
- A comparison against measured temperatures (1880 to 2019)
- A comparison against IPCC future predictions
- A comparison with other more sophisticated models

The Dependency on IPCC Work

Whilst the formulation of the model is independently derived, it uses a range of parameters reported in IPCC documents. So, the first question is, 'How uncertain are these IPCC parameters?' This was covered in the Global Warming Handbook. and is summarised in Table 14.1.

TABLE 14.1 Uncertainties in IPCC Reported Work

Parameter	Uncertainty 2 σ	Modelling Comment
Measured Temperature	≈±0.2°C	Needed to validate models
Greenhouse Gas Forcing	roughly ±10% for ≈1 °C change	Good enough to validate the effect
Aerosol Effect Important for Understanding	±50%	Needed to predict ahead. See Figure7.5 AR6
Climate Feedback Vital for understanding	≈50%	Vital to predict ahead.
Ozone	≈50%	A significant contributor
Energy Budget Heat Imbalance	±30%	Important for present and future

It is important to realise that uncertainty in modelling global temperatures is significant. and particularly so in several areas (aerosols Atmospheric Feedback, Global Energy budget, and ozone).

Use is made of 'greenhouse gas Radiative Forcing' in our model and this represents the best available and most certain of the available data. We also make use of the IPCC figures for aerosol forcing, and despite its uncertainty, using its best estimate value enables this, and no doubt other models, to track measured global temperatures.

In the area of atmospheric feedback, our model is not based on IPCC-reported approaches. It stands alone and only relies on measured data, notably sea surface (water to air) temperature differential and NASA satellite-based power fluxes. This is a useful estimate in an area that has often been targeted for criticism.

At this point we need to be realistic about IPCC calculations of future temperatures. They have significant uncertainties. In essence they are 'best estimate' projections – a best attempt to reconcile uncertainties.

Let's have a look at how our model compares with measured temperature records and IPCC models.

Model Comparisons

Table 14.2 is a comparison table. The first entry compares our model with the actual measured temperature record. It is a numerical comparison of what we have already seen in graphical form. The mean deviation between our model and the temperature record (using 20 data points) is 0.002 °C. This reflects the tendency of the model to lie in the middle of the measured data (no fiddle factors are deployed). Also, the table shows that the Standard Deviation of the model from measured data is about 0.1°C. This means that we are 68% confident that the model reflects the measured data within ±0.1 °C and 95% confident. that the model reflects the measured data to ±0.2 °C. This aligns with a similar figure for IPCC assessments. Additionally, the second entry in the table shows that the model predicts the accepted baseline temperature of 255K in the absence of greenhouse gasses (including water vapour). This bounding case supports the conclusion that our illustrative model is sensibly based.

The remaining entries compare this model with IPCC reports. Agreement with IPCC modelling reflects independent agreement with worldwide scientific effort. In all cases our model's performance parameters lie well within the uncertainty bounds for the mean values for the 20 or so IPCC models. I am not going to run through them all. However, I will highlight the area of atmospheric feedback. In this instance, the comparatively simple analysis that underpins our model produces a result that (in terms of feedback) is in numerical agreement with the more complex IPCC work.

Finally, our model is within 7% of the 2018 'National Plans' forward estimation for 2100, providing a final measure of reassurance. However, I suspect that both estimates essentially indicate a change exceeding 3 °C.

In conclusion, the comparisons above suggest that our model is more than adequate as an illustration of the physical mechanisms at work.

Table 14.2 Model Performance Comparison

	This Model	IPCC uncertainty
Agreement with Measurements from 1880	0.22 °C 2 σ (to end of 2020)	≈0.19 °C AR5 2 σ
Surface Temperature No Water Vapour. No CO_2	255 K	255 K
CO_2 Radiative Forcing (RF) RF at 2011 (391.85 ppm CO_2) RF & ERF 2019 411.66 ppm CO_2 All w.r.t. pre-industrial 1750	RF 1.81 W/m² RF 2.07 W/m²	RF 1.82 W/m² ERF 2.16 W/m²

All WMGHG RF
   RF 2011
   RF & ERF 2019
	2.85 W/m²
	3.26 W/m² (RF)
		2.83 W/m²
		3.32 W/m² (ERF)

Doubling of CO_2 (RF) 399.57 to 799.14 ppm	RF 3.63 W/m²	AR5 RF 3.7 W/m²
Transient Climate Response Estimated from Model	2 °C	1.8 °C (1.2 - 2.4) AR6
Net Climate Feedback – overall model	1.1 W/m²/°C Best Estimate	1.16 W/m²/°C (±0.5)
Water Vapour Feedback (an independent estimate)	1.84 W/m²/°C Best Estimate from Model	1.81 W/m²/°C (±0.2) AR6 sect. 7.4.2.2 combined estimates

(σ is the Standard Deviation)

New Areas of Inclusion and Areas of Omission

In this model, I have included the warming effect of Ozone (Section 11) as well as the damping effect of the oceans in Section 12. This produces a much more comprehensive model and covers all the significant issues covered in Chapter 7 of the 2021 IPCC report, AR6.

I have also explained that my treatment of water vapour feedback shows a decline with increasing atmospheric temperature and I have omitted a correction for this. I estimate this effect as a few hundredths of a degree.

Notes:
- *Standard Deviation and Mean Deviation are different ways of working out the agreement of the model with real measured temperatures. Standard deviation is more appropriate when the data varies cyclically around a central tendency. Also note that, for clarity, the model used in this book only uses a subset of measured data.*

Further Information:

IPCC, AR5 and AR6, The Physical Science Basis.

Both are available online via the IPCC website.

15 - Global Warming Spreadsheet

I am aware that this book may have differing types of readers, with many different uses in mind: from those who want to learn, those who want to educate others and those who have an enquiring and scientific mind. Some may want to build their own climate models and want a starting point. Others may just want to run this model - and to do this, they want an appreciation of the physics behind what they are doing. To support these aims, I have provided a spreadsheet containing the model described here. This is broken down into several linked worksheets, which I have created and used myself to write this series of books.

This spreadsheet is available to download via the *http://www.Earth2hot.com* website link. It uses the commonly available Microsoft Excel program.

I am afraid that am not in the business of creating, selling, and maintaining spreadsheets. My spreadsheet is available for free – as an aid to understanding and is not sold to you as part of your book purchase. Please bear in mind that it is a demonstration – not a refined software product. After you have downloaded the spreadsheet, I would recommend that you immediately make a copy as a backup.

The spreadsheet is broken down into several worksheets. They tend to follow an explanatory progression, and Appendix 8 gives a brief explanation of each worksheet. Within each you will recognise the model described in this book - translated into accessible calculations. Although one of them, the Auto Chart worksheet, will probably take up the most of your attention, as you can simply 'dial in or out' the warming processes described in this book. This enables you to immediately view the impact in graphical form. You can then start on a journey of your own: to explore global warming.

16 - Final Thoughts

My initial work with a simple model persuaded me that illustrative modelling of Global Warming is achievable in an approachable manner.

I have developed this model further to establish (initially to myself) that human activities have caused Global Warming. The preceding book establishes the link between warming and human activities and makes use of this model as a significant part of the evidence.

This is of primary importance – because discussing Climate Change in the absence of certainty in a warming process is meaningless. It is also important to do this without diverting into climate changes (sea level changes, glacial retreat etc..) as a means of establishing warming – because none of them establish the cause.

Already, this simple model has provided insights for me:
- That a small amount of CO_2 causes significant changes in Global Temperature.
- That CO_2 is just one of many gasses that significantly contribute to Global Warming
- That industrial and volcanic aerosols cool the planet and may form a basis for contingency action
- That temperature-based Water Vapour Feedback is significant and can be simply approximated
- **That the induced water vapour changes in the atmosphere need to be highlighted as a primary climate concern**
- **That the oceans are perhaps our greatest saviour, absorbing large amounts of heat over time**

However, if you have read this and the preceding book, you should have concluded that even the most sophisticated assessments of future Global Temperatures are 'best-estimates'. Should you use this model or one like it, you can enter the world of Global Warming and make your own best estimate assessments.

As an outsider in this debate, I have realised that there is a valuable use for a simple physically based model, one that illustrates the significant aspects of Global Warming. By making this viewpoint as approachable as I can, I hope that I have contributed to demystifying this subject and encouraged others to get involved.

Importantly, using a model gives you the ability to modify it and gain insights. It offers the possibility of asking 'What if' questions and testing the outcome.

Given the politics surrounding this topic, it is then possible to separate reality from fiction and exaggeration from ignorance. You can now judge for yourselves.

Jim Wiles - January 2022 (Earth2hot@gmail.com)

Postscript

The Global Warming Handbook will be updated once more to bring it in line with this book. I will, from time to time endeavour to update the Excel Model, available for free on http://www.Earth2hot.com, *and post relevant change notes alongside this.*

Figure 16.1 Climate Monitoring Buoys
- a lonely presence

Other Books in the Series

Global Warming Handbook
By Jim Wiles

The Global Warming Handbook tries to make things simple. Whilst being true to what the scientists say, it asks and answers key questions with words and pictures:

How does our world get warm?
Is Carbon Dioxide the main cause?
What else causes global warming and what is the true cause?
How uncertain is the science?
What affects the timing of warming?
When should we panic?
What are tipping points and is it is it easy to deal with them?

All in a short book.

Simulation Made Easy
By Jim Wiles

How to produce **Dynamic** Models the Easy Way, with step-by-step graded examples - from basic mathematical functions to the simple physics of a pendulum, and ultimately to Engineering systems, **Global Warming** and even Covid 19.

Illustrated using a General-Purpose Simulation Program (SimAzing), which is free to download. (See http://www.simazing.com/).

The Author

Jim Wiles is a physicist with extensive experience as a lead author, working with teams of scientists to draw together complex scientific projects. Dealing with complex science and reducing it to core issues, together with a background in applying computers, gave him the opportunity to simplify the complexities of Global Warming.

He has also walked the debris-strewn paths of receding glaciers.

Acknowledgements

Images Under Licence from Adobe Stock

Figure 16.1 Lonely Climate Monitoring Buoys,52521143, <Dabarti>

Public Domain Images (PD) and Licenced as Creative Commons 2.0

Figure 1.1 Maintaining a Climate Monitoring Buoy, NOAA Photo Library - Linda Stratton 2003 (PD)

Figure 5.1 Cumulonimbus, NOAA / AOML / Hurricane Research Division, NOAA photo Library fly00890 (PD)

Figure 7.3 Summit Supercomputer 2018, Carlos Jones / Oak Ridge National Laboratory (CC-by-2.0)

Figure 10.2 A Coke Production Plant in 1942, US Farm Security Administration, Smokestacks Alfred Palmer (PD)

Figure A6.2 Electromagnetic Absorption by Water Vapour – John Bertie

Figure A6.1 Earth's Heat Balance, NASA

Section 12 Notes JPSS Satellite Image, Author NOAA Satellites

Appendix 1 - Absorption by CO_2

Any infrared radiation from the earth's surface that is not absorbed will radiate to space directly. It will follow the standard laws of physics.

$$\text{WinPower} = K_3 \cdot \sigma \cdot A_e \cdot T_s^4$$

Where Ts is the surface temperature

A_e is the earth's surface area

σ is Stefan's constant

K_3 is a radiation transmission factor

Under ideal circumstances, **K_3** would be 1. In other words, 100% of the possible power, based on surface temperature, would be radiated into space. But we know that greenhouse gasses absorb thermal radiation and only a fraction can escape directly. This means that the overall value, **K_3**, allows for this absorption by all greenhouse gasses, as well as surface emissivity. (A value of 1 is used for emissivity in the model we are creating.)

NASA gives a figure for the contemporary amount of power radiated directly of about 40 W/m² and explains the role played by satellites in working out a global heat balance. At a given surface temperature, **K3** can be found from:

$$40 \ (W/m^2) = K_3 \cdot \sigma \cdot A_e \cdot T_s^4 / A_e$$

Therefore:

$$K3 = 40/(\sigma \cdot T_s^4)$$

I chose 2005 for the reference year, as it was near the time when I first looked into global warming and it was the point where the rise in global temperature became statistically significant. At this date, we know our reference values for the CO_2 level (C_0 used further down) and the temperature (Tsref). With **Tsref** =287.77, and $\sigma = 5.67 \cdot 10^{-8}$

K_3 was evaluated as 0.10286 and is dimensionless.

K_3 = Transmission = (1 − absorption) and therefore

Absorption = (1 − Transmission) = (1 − K_3) = 0.8971

The overall reference level of absorption, Aref, is 0.8971 (at 2005)

We need to derive other values from our reference conditions. Specifically, we need to split absorption into two terms: 'a_{1ref}' essentially for water vapour and 'a_{2ref}' for CO_2.

$Aref = (a_{1ref} + a_{2ref})$

$WinPower = (1 - (a_{1ref} + a_{2ref})) \cdot \sigma \cdot A_e \cdot T^4$ (W)

If we know the proportion of the greenhouse gas effect due to CO_2 compared to 'water vapour and other greenhouse gases' we can calculate values for a_{1ref} and a_{2ref}.

In the equation above, the coefficient for absorption from water vapour, a_1, dominates. Also, at this point, **a1 is assumed sensibly constant with a value of a1ref (later,** we will separate out the absorption of other greenhouse gasses). From the above, we can see that if a_{2ref} (the absorption due to CO2) increases to some value a_2, then the radiated power reduces. We also know that a_2 changes with CO_2 concentration and various references refer to the relationship between CO_2 and absorption as logarithmic.

If we know the measured CO_2 level (C_0) a general equation can be is used to describe this variation of absorption with CO_2 level, '**C**' in ppm.

$a_2 = a_{2ref} \cdot [1 + \ln(C/C_0)]$

We can now calculate the value of absorption for a given value of CO_2 level. Our complete equation for the power transmitted through the thermal atmospheric window at any surface temperature and CO_2 level now becomes.

Winpower = $\{1 - a_{1ref} - a_{2ref} \cdot [1 + \ln(C/C_0)]\} \cdot \sigma \cdot A_e \cdot T_s^4$

Note again that when $C=C_0$, our value for absorption is $a_{1ref} + a_{2ref}$ or Aref, as noted above.

We can also use this relationship to estimate the conditions in, say, 1890 and use this as a new baseline, or reference for comparison with temperature changes since then. Temperature measurement data from monitoring stations exists from 1880, but I have used 1890 as a reference to maintain consistency with later feedback calculations.

Notes:

When deciding on a reference level of CO_2 this will be at a specific date, for a measured temperature and measured CO_2 level.

The proportion of CO2 absorption as a percentage of Water Vapour absorption is needed to complete this analysis. An example is given in the main text (section 4), and this is evaluated further in Appendix 2

Appendix 2 - The IPCC Forcing Relationship

IPCC reports state that the relationship between Radiative Forcing (W/m^2) and CO_2 (ppm) is logarithmic. In this context, Radiative Forcing is the power (flux in W/m^2) radiated into space. Now Radiative Forcing is an unusual parameter. It is defined as the instantaneous power change (per square metre), at the top of the troposphere, induced by a change in some variable, e.g., CO_2. Its primary purpose is as a comparison guide between a range of drivers that increase or decrease global warming. The logic behind this is to enable the comparison between, say, a change in arctic ice cover, with a change in methane level, or a change in CO_2 level. However, in modelling terms it represents an instantaneous transient and is very useful in modelling terms. Not surprisingly, this did not meet the IPCC goal of achieving some simple comparative and additive parameter, and since AR5 this has morphed into an Effective Radiative Forcing (ERF), which embraces instantaneously acting feedback terms. Our model estimates the effect of feedback for all changes in forcing as a separate and later stage in our model. Therefore, ERF is not the best parameter when defining our model.

In consequence we use RF wherever possible in constructing our model. It requires that the surface temperature remains constant whilst this change is calculated and is often expressed in relation to an appropriate reference level of pre-industrial times. In the context of greenhouse gasses, it is provided from absorption 'codes' taking into effect atmospheric conditions. This is often referred to as the 'kernel' in IPCC reports. From this, the changes in RF are calculated using simplified relationships.

Specifically, for CO_2, an IPCC relationship is stated as a change in Forcing (Watts/m^2) for a change in carbon dioxide (ppm).

$$\Delta F = 5.35 \ln (C/C_0)$$

Where **C_0** is the reference CO_2 (ppm) value at a reference date or period and **C** is the CO_2 concentration at some other time. The term, 5.35, is derived from radiative transfer modelling – This considers a mixture of cloudy and clear sky conditions at several locations. This is the equivalent of:

$$\Delta F = \ln [(C/C_0)^{5.35}]$$

Which seems an unusual expression to describe a marginal effect. In that the dependency on C/C_0 has an exponential component.

Moving on - in forming the model used in this book, we are interested in the change of absorption for a change in CO_2. From which we can work out a power change and finally a surface temperature change. We know that power escaping from the earth's surface must be governed by the laws of radiation. From Section 4

Power Escaping = (1-absorption) $\cdot \sigma \cdot A_e \cdot T_s^4$

Hence Radiative Flux 'F' is obtained by dividing by A_e

$F = (1\text{-absorption}) \cdot \sigma \cdot T_s^4 \quad W/m^2$

The IPCC *change* of forcing is termed ΔF and is an imbalance, a change in flux. However, as absorption increases so Forcing, a measure of warming imbalance. will increase. Inspecting the equation above we have:

$-\Delta F = -\Delta \text{absorption} \; \sigma \cdot T_s 4$ (a positive change)

σ is a constant, and T_s^4 is held constant by the definition of forcing. This applies at any value of T_s, meaning that the simple IPCC logarithmic forcing relationship also applies to absorption. Moving on, we can state.

$5.35 \; Ln(C/C_o) = \Delta \text{absorption} \; \sigma \cdot T_s^4$

From Section 4 and Appendix 1, CO_2 absorption, in isolation, can be expressed in the general form:

$a_2 = a_{2ref} \cdot [1 + \ln(C/C_0)]$

Therefore, the change in a_2 is:

$\Delta a_2 = a_{2ref} \cdot \ln(C/C_0) = \Delta \text{absorption}$

The value of a_{2ref} can be evaluated using the IPCC forcing relationship:

$5.35 \cdot \ln(C/C0) = -\Delta \text{absorption} \cdot \sigma \cdot T_s^4$

$5.35 \cdot \ln(C/C_0) = - a_{2ref} \cdot \ln(C/C_0) \cdot \sigma \cdot T_s^4$

Therefore, using a reference year of 2005 as noted in the main text ($T_s = 287.77$):

$a_{2ref} = 5.35 \cdot \ln(C/C_0) / (\ln(C/C_0) \cdot \sigma \cdot Ts^4)$

$a_{2ref} = 5.35 / (\sigma \cdot Ts^4)$

$a_{2ref} = 5.35 / (5.67 \cdot 10^{-8} \cdot (287.77)^4)$

$a_{2ref} = 0.01376$

Now:

$Aref = a_1ref + a_2ref \cdot [1+\ln(C/C0)]$

(CO_2 plus water vapour etc.)

For a reference level of CO_2, knowing that when $C = C_0$

$0.8971 = (a_{1ref} + a_{2ref})$ (Overall Ref. level from Appedix1)

$0.8971 = (a1ref + 0.01376)$

$a1ref = 0.8971 - 0.01376 = 0.8833$

Also, the reference relative absorption worth of CO_2 is

0.01376/0.8971, which is 1.53% of the total absorption by other greenhouse gasses (water vapour, methane, **etc.**)

Appendix 3 - IPCC Forcing Relationship Validity

If we want to predict ahead, we need to be sure that we are using an absorption relationship that is valid for the future range of CO_2 values. Just to recap, we use a logarithmic relationship in our modelling of power radiated though the thermal Atmospheric Window.

$Winpower = [1 - a_{1ref} - a_2] . \sigma . A_e . T_s^4$

Where:

$a_2 = a_{2ref} . [1 + \ln(C/C_0)]$

The relationship that is used in this book is based upon the IPCC Forcing relationship noted in Section 7:

$\Delta F = 5.35 \ln (C/C_0)$

(1998 revision used by IPCC)

But where does this come from? What is it based on? It is clear that the constant 5.35 is based on radiative transfer modelling for three atmospheric profiles: 'Equatorial', 'non-equatorial Northern Hemisphere', and 'non-equatorial Southern Hemisphere'. It utilised a range of radiative transfer models and appears comprehensive. However, it does not explain the logarithmic relationship. The basis for this reaches further back in time. Working the trail backwards, the equation above was a revision of a previous equation:

$\Delta F = 6.3 \ln (C/C_0)$

(1990 IPCC AR1 Scientific Assessment of Climate Change
table 2.2. - stated as valid for CO2<1000ppm

Again, the functional form predates this – and the trail becomes more difficult and goes back to a publication in Climate Monitor, an in-house publication by the University of East Anglia Climate Research Unit. Thanks to research by Stephen McIntyre on Climateaudit.org, we arrive at:

$\Delta F = 6.33 \ln (C/C_0)$

1987 Wigley, based on work in the range 250ppm to 600ppm

This latter work explains that the appropriate relationship is linear at low concentrations, square root at intermediate values and logarithmic at higher concentrations. That's absolutely fine, but it seems clear that the overall relationship does not appear to be based on a physical analysis. It is a numerical fit to the results of modelling. This still raises the question: 'over what range is it valid'? The only definitive statement is that it is valid from 250 ppm to 600 ppm CO_2 .– and like the reference to <1000 ppm in the first IPCC report, it contains no background explanation of these limits.

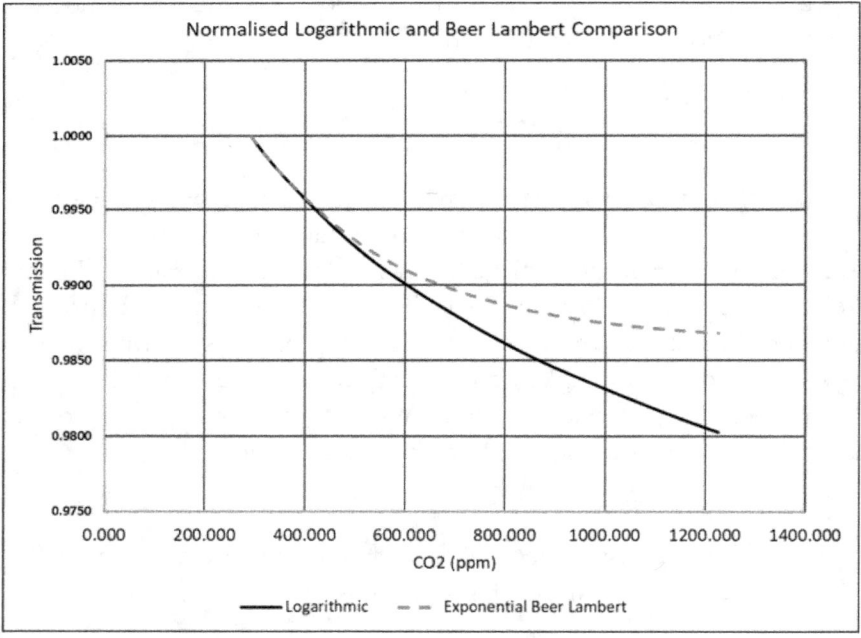

Figure A3.1

There is something called the Beer-Lambert law, which is used by chemists to evaluate the concentration of constituents in a fluid or gas. This is done by measuring the transmission of light through a fluid. The lower the transmission the higher the absorbing content in the fluid. This can be turned around to work out the absorption from a knowledge of the concentration. The transmission, Tx, compared to some reference value, Tx_0, is given by:

$$Tx = Tx_0 \cdot (1 - e^{-(C/C_0)})$$

Figure A3.1 provides a normalised comparison between the IPCC logarithmic relationship and Beer-Lambert, using a reference level for a_2 at 1890. You can see that both approaches yield virtually the same result up to CO_2 concentration levels of around 450 to 500 ppm (the current levels for 2019 are around 414 ppm.) After that, the IPCC logarithmic approach shows a more detrimental loss of transmission (more global warming) whilst Beer-Lambert saturates (the warming effect of CO_2 effectively stagnates). Surprisingly, the IPCC marginal effect of residual CO_2 absorption appears to have a continuing and greater magnitude than if treated as a primary effect. Clearly, feedback effects may worsen any warming (note again that I deal with these separately to expose the physics involved).

The bottom line is that I found it difficult to identify the limitations of the IPCC relationship.

The recent (September 2021) IPCC AR6 report now contains a few lines summarising the background and points to a revised simplifying expression for **ERF**. Looking at the reference documents feeding into AR6, we can see that these relationships have validity backwards in time, some thousands of years. Unfortunately, I have not been able to locate a clear statement as to whether this validates the expressions into the future.

Appendix 4 – Changing the Model Reference Date

In several instances, it is desirable to change the reference date used by the model. The reason for this is that often IPCC data will be based on the difference between pre-industrial forcing levels and the forcing levels close to the date of their publication.

In Appendix 2, IPCC forcing data for the effect of CO_2 was converted to absorption. A reference date of 2005 was used – because information on both forcing and thermal Atmospheric Window flux was available at this date. Also, a statistically significant rise in global temperatures was confirmed by this date. However, using this 2005 reference point would have left the resulting graph floating in mid-air. The result is shown in Figure A4.1.

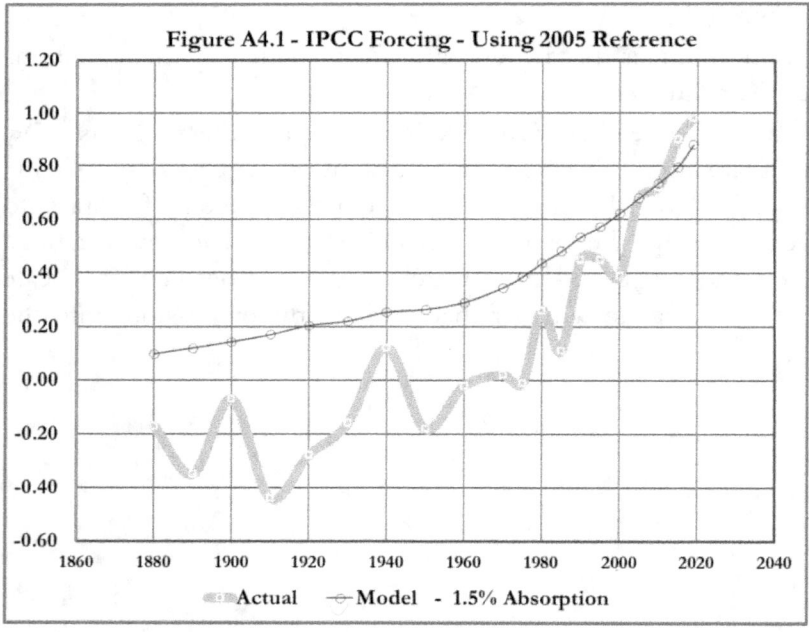

It is easy with our model to sort this out. All we need to do is, using the 2005 referenced model, identify the thermal atmospheric window radiation for, in this example, 1890. Then reset the initial conditions to the measured temperature and CO_2 level in 1880, together with the revised thermal window radiation figure. The result is shown in Figure 7.1 repeated below.

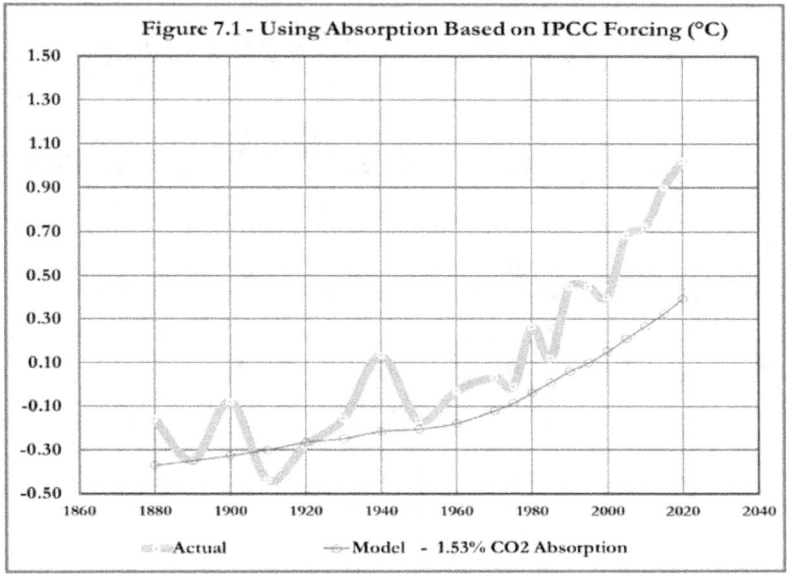

Spreadsheet reference data for the simulation graphs used in this book are shown below.

CO$_2$ absorption - Available Reference Conditions - these apply to all Worksheets						
		Ts (K)	CO$_2$ (ppm)	Thermal WinFlux (W/m^2)	Relative Absorption (CO$_2$ %)	Albedo
Preset Conditions						
1 ○	Try 3.62% 1890 ref	286.80	292.46	42.48	3.620	0.3000
2 ○	Try 2.77% 1890 ref	286.80	292.46	41.85	2.770	0.3000
3 ○	IPCC CO2 Only 1890 Reference	286.80	292.46	41.10	1.530	0.3000
4 ◉	All WMGHG 1890 Reference	286.80	292.46	41.64	2.665	0.3000
Initial Reference examples						
5 ○	IPCC CO2 only 2005 Ref.	287.83	378.90	40.00	1.530	0.3000
6 ○	All WMGHG 2005 Ref.	287.83	378.90	40.00	2.371	0.3000
User Values						
7 ○	User Values 1					
8 ○	User Values 2					
9 ○	User Values 3					
	You can edit the User Values and then select them					

Appendix 5 – Evaporative Feedback – the Balance of Evaporation

Following on from the explanation in Section 10, the solution to the balance of evaporation at ocean surfaces is in three stages.

(1) Evaluate the Reduction in Radiative Flux Due to Evaporation

To resolve Water Vapour Feedback, we need to consider surface thermal radiation via the thermal Atmospheric window (**Radiative Flux**) but also how evaporation occurs in the boundary layer just above the ocean surface. This approach to water vapour feedback is only possible because TAO buoy measurements bypass the need to atmospheric simulation.

Let's look at the boundary layer more closely.

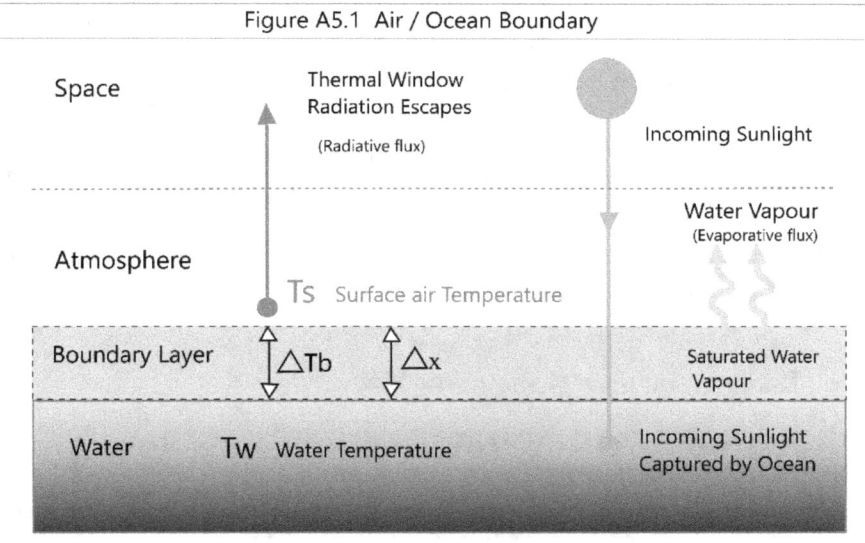

Figure A5.1 shows the air boundary layer of the ocean. The average difference between ocean surface water and air temperature (**ΔTb**) has been evaluated using climate monitoring buoy measurements as 0.82°C. This is for Pacific Ocean buoy data reviewed by Bob Tisdale covering the period from 1995 to 2010. Covering the tropical zone (roughly 10 degrees North and South of the equator).

For the moment, let's examine this situation as a simple thermal boundary layer. In Figure A5.1, the thickness of the boundary layer is designated as **Δx**. This dimension can be evaluated by assuming conduction dominates a thin boundary layer above the ocean and using known measurements of air and ocean temperatures.

In particular, we need the power transmitted through the boundary to evaporate water. I refer to this in this appendix as the **Evaporative Flux**. This is designated as F_{LH} (as noted in previous sections of the main text) and has been evaluated by NASA as 86.4 W/m².

$F_{LH} = K \cdot (\Delta Tb) / \Delta x = 86.4$ W/m²

Using:

K, the thermal conductivity of air, of 0.0255 W/m²/°C

$F_{LH} = 86.4$ W/m² and $\Delta T = 0.82$ °C

$\Delta x = K \cdot \Delta Tb / F_{LH}$ (metres)

Δx = 0.24mm (240 microns) ....(for information only)

The important point is that, undisputedly, evaporation from the ocean occurs in such a boundary layer.

Without such a layer, radiation would occur at **Tw**, the ocean surface temperature and its radiative power would be:

Radiated Power = $Tgh \cdot \sigma \cdot A \cdot Tw^4$ W

Where 'A' is the surface area
and Tgh is an absorptive transmission factor.

Using the term **Radiative Flux**, which is the power per unit area radiated from the surface:

Radiative Flux(Tw) = $Tgh \cdot \sigma \cdot Tw^4$ W/m²

with a boundary layer this becomes:

Radiative Flux(Ts) = $Tgh \cdot \sigma \cdot Ts^4$ W/m²

You may have noted that I have used **Ts** here, which is the variable name used within this book for air surface temperature, typically measured at a nominal height of 1.5m. In doing this, I am assuming that Ts is essentially the same as the temperature immediately above the boundary layer (I will return to this in due course). At a later stage I will explain that the global average surface temperature can be used in place of the air temperature above the ocean.

Measurements are available for **Tw** and **Ts** and we can easily work out the reduction in power that is radiated. The proportional reduction in radiative power flux is:

$$Prf = Ts^4/Tw^4 \quad \text{dimensionless}$$

It is clear that as we are no longer radiating as much heat directly into space, then **our planet must warm**. This heat is effectively being used to evaporate water - adding heat to the atmosphere. As it stands, **Prf** is the steady state reduction in radiated power flux at a point in time. We need to know the change in **Prf** with each ("greenhouse-gas" induced) warming increase. Now, **WinFlux** is the power flux radiated via the thermal atmospheric window (assessed by NASA as $40W/m^2$ for contemporary times), and we can calculate the radiative flux reduction **Rf**.

$$Rf = (Ts^4/Tw^4 - 1) \times WinFlux$$

We now know that a reduction in radiative power depends on the boundary conditions (**Ts**) and ocean surface water temperature (**Tw determined from the known ΔTb**), as well as WinFlux.

Our prime interest here is the change in **Rf** with surface temperature **Ts**. This is because we model the global average **Ts** (dominated by the ocean temperatures), and we know how it changes with greenhouse gas concentration from measurements. Using some simple calculus:

$$dRf/dTs = (4Ts^3/Tw^4) \times WinFlux \quad \ldots\ldots \text{ \textbf{Equation (A}}_{5.1}\text{\textbf{)}}$$

In this analysis, Tw can be estimated (using the known **ΔTb) – Ts** and **Winflux**, at historic levels of CO_2, are already available from our model. Therefore, this equation defines how radiative losses change with changes in measured global mean air temperature.

(2) The effect of ΔRf on Evaporative Flux

We can now move a little forward in working out water vapour feedback. Crucial to this is recognising that the change in radiative flux via the thermal Atmospheric window must translate to the change in evaporative flux F_{LH}, as noted above. (Assuming that the reduction in radiation results in an increase in evaporation (and evaporative flux).

From the conservation of energy, we immediately know a key parameter, the proportional change in evaporative flux.

$$\Delta F_{LH} / F_{LH} = \Delta Rf / F_{LH}$$

As noted above, NASA quote the contemporary value of F_{LH} as 86.4 W/m², which presumable takes into account evaporation globally and includes locations where evaporation is lower than the tropics.

(3) The Change in Water Vapour Density with Temperature

We now need to consider the role of humidity. Specifically, the mass of water vapour that can be held in a cubic metre of air. In fact, it is the maximum amount of water vapour - the saturation vapour density that is of interest.

As we a creating a numerical model, we can use a polynomial to evaluate the amount of water vapour present in the atmosphere under saturation conditions. This saturated state is what we can expect just above the ocean surface, within a boundary layer. The polynomial used is:

Vapour Density $= 5.018 + 0.32321 \cdot T + 8.1847 \times 10^{-2} \cdot T^2 + 3.1243 \times 10^{-4} \cdot T^3$

..................(in gm/m³)   (for source see Appendix 5 notes)

It is useful to look at an extract from a table generated using this polynomial.

Each entry in Table A5.1 shows a deliberately random progressive increases in temperature and calculated mass of water vapour (**m**) in kg/m². This is similar to the layout in a global warming spreadsheet shown in the main text (Figure 10.1). Although, in the case of global warming, the temperature changes are due to greenhouse gas concentrations. We can use such a table to calculate the changes in water vapour mass, **Δm** (kg/m³), as we move from one temperature to the next by simple subtraction.

Now, the mass of vapour and its change are all directly related to radiative power changes – we need heat to achieve the vaporisation of any given mass of water. This type of table can also be used to evaluate the ratio **Δm/m**, the proportional change in latent heat flux (evaporative flux), and we can also directly calculate **Δm/m** per °C change in air temperature (**Ts**).

Vapour Density Changes with Temperature Table A5.1

Ts	Vapour Density	Temperature Increment	Change in m	Relative Change	Relative Change per °C
	m	ΔT	Δm	Δm/m	Δm/m/ΔT
(°C)	(gm/m³)	(°C)	(gm/m³)		(per °C)
15.50	13.158				
15.62	13.257	0.123	0.0993	0.00749	0.061
15.82	13.414	0.194	0.1575	0.01174	0.061
15.94	13.518	0.126	0.1033	0.00764	0.061
16.12	13.668	0.181	0.1499	0.01097	0.061
16.25	13.769	0.121	0.1010	0.00734	0.061
16.41	13.908	0.165	0.1391	0.01000	0.061
16.54	14.015	0.126	0.1070	0.00764	0.061
16.66	14.122	0.126	0.1074	0.00760	0.061
16.84	14.277	0.180	0.1547	0.01083	0.061
16.96	14.375	0.113	0.0982	0.00683	0.061
17.06	14.464	0.102	0.0892	0.00617	0.061

We can now synthesise a relationship between our change in temperature, increasing mass of water vapour, and radiative loss for a given change in temperature and saturation mass.

$\Delta Rf/F_{LH} = K_m \times (\Delta m/m) + $ constant

The constant must be zero (when $\Delta m = 0$, $\Delta Rf = 0$). Therefore:

$\Delta Rf/F_{LH} = K_m \times (\Delta m/m)$

$$\Delta Rf = K_m \times (\Delta m/m) \times F_{LH} \quad \ldots\ldots\text{Equation A}_{5.2}$$

We are used to looking at equations with an objective of solving them for a specific value, but they do in a broader sense express a balance. At this stage we need to recognise that **ΔRf** is one side of such a balance and from (1) above we know it represents the boundary layer conditions in terms of radiation. On the other side we have an expression related to evaporative power. They are related by a simple co-efficient **Km**. Working backwards; in terms of feedback, we need to recognise that **ΔRf** is the required change in radiative flux for a given change in the mass of water vapour. If we evaluate **Km** and multiply it by **Δm/m** and **F_LH** we know, for a thermally balanced boundary layer, how much the radiative term must change for a change in air temperature – how much feedback is involved.

To work out **Km** we need to make use of a thermodynamic property of saturation density. Look again at Table A5.1 and inspect the column headed **Δm/m per °C**. As temperature increases and we progress down the table we observe a nearly constant value of 0.061 per °C change in air temperature. There is a slight downward variation when we look at a wider range of temperature variation, but for contemporary Global Warming purposes, **Δm/m per °C** is essentially constant. Remembering that I chose a random progression of temperature for this table, we can see that this is not an artefact of global warming but a physical property.

Making use of this solely for the purposes of evaluating Km, we can restate equation (A$_{5.2}$) in terms of the change in water surface temperature **ΔTs**.

$$\Delta Rf/\Delta Ts = K_m \times F_{LH} \times ((\Delta m/m)/\Delta Ts)$$

Making use of equation (A$_{5.1}$) which is restated below:

$$dRf/dTs = (4Ts^3/Tw^4) \times \text{WinFlux}$$

$$(4Ts3/Tw4) \times \text{WinFlux} = K_m \times F_{LH} \times ((\Delta m/m)/\Delta Ts)$$

......for small changes in Ts and Δm .... Yielding:

$$(4Ts^3/Tw^4) \times (\text{WinFlux}/F_{LH}) / ((\Delta m/m)/\Delta Ts) = K_m$$

As $((\Delta m/m)/\Delta Ts) = 0.0610$ see above

$$(4T_s^3/T_w^4) \times (\text{WinFlux}/F_{LH}) / (0.0610) = K_m$$

Equation (A5.3)

Which is dimensionless

Under contemporary conditions, we can now work out a value for **Km** for:

(T_w = 15.5 °C = 288.65 K, ΔT=0.82 K, T_s = 287.83K, Winflux=40W/m², F_{LH} = 86.4 W/m²

Km = 0.104   dimensionless

For our model, we can now make use of Equation (A5.2) within our model to evaluate the amount of feedback for a change in evaporated mass:

$$\Delta R_f = K_m \times F_{LH} \times (\Delta m/m)$$

In our model, I used the polynomial above to calculate values of ($\Delta m/m$) for the temperatures due to global warming for different levels of CO_2. This enables the calculation of the change in thermal radiation due to evaporation, and this can be applied as feedback via thermal Atmospheric Window radiation – which will, in turn, result in an amplifying change in global warming.

The feedback in terms of **mass of water vapour change** is

Feedback = 9.01 W/m² for changes in ($\Delta m/m$)

-under contemporary conditions.

(Note, rounding errors cause a negligible difference between this and the main text value).

This value of feedback is obtained using the contemporary 86.4W/m² evaporative flux (evaluated from satellite measurements). This is the average flux for the planet. Evaluating for both tropical and high latitude conditions and then working out a mean effect produced a value within 5%, suggesting that we can use this as a general feedback value.

The model applies this feedback as it iterates to a solution. This means that each iterative change due to water vapour will also generate a further change and so on - I was pleased when it converged and thermal runaway did not occur!

Figure A5.2 shows the effect of evaporative feedback for the time period up to the end of 2019. When we apply this feedback term, we significantly increase our agreement with measured temperatures.

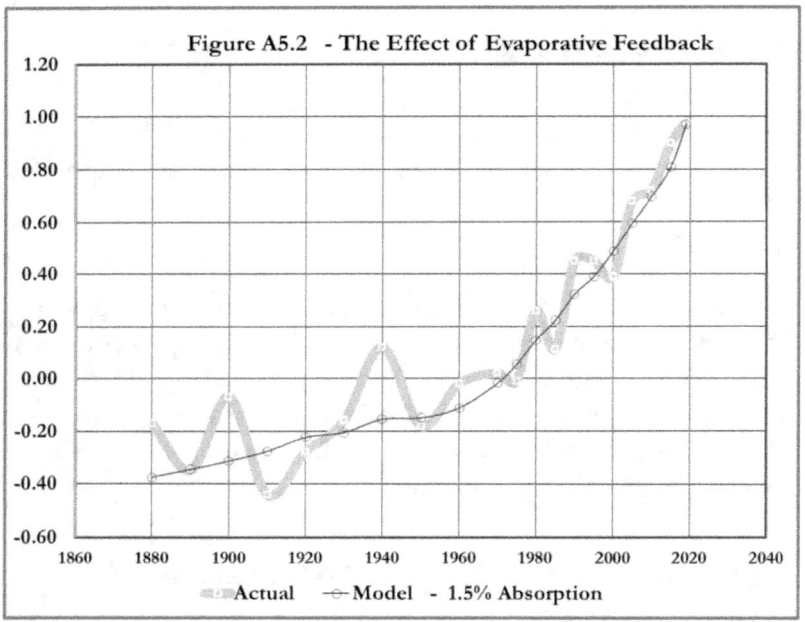

There are a few issues left dangling in this analysis, and I will try to mop these up. I am suggesting using the feedback coefficient above as a constant. From the equivalent changes from 1890 to 2018 the calculated feedback term at 1890 increased by +4 %. Trials with the model showed this to be equivalent to an average surface temperature change in terms of a few thousandths of a degree. Similarly, in terms of projecting forward to 2100, where Km is reduced by −11%, this resulted in a 0.06°C reduction in predicted temperature. Incidentally, these estimates do show that the Water Vapour Feedback effect reduces as global warming increases. It also suggests that we could apply this as a correction that is not currently included in the model. In terms of an illustrative model that respects basic physics, the variation of 'km' does not seem to be an issue.

The Air Temperature used in the analysis

As a heat transport mechanism, global evaporation predominates in the tropical zone – but not necessarily at the equator. This is because wind speeds are low to non-existent at the equator and increase as you move North and South in the tropical zone. Further, purely equatorial evaporation can be straight up and down, with high humidity and torrential rain. Now, despite the overall dominance of the tropical region, evaporation can occur anywhere where exposed water is present. This means that it cannot be appropriate to use purely tropical or ocean temperature conditions in the model. However, because of the way in which the global mean temperature is evaluated, it is dominated by ocean temperatures (mean 16.1 °C) over the land temperatures (mean 8.5 °C) in forming the twentieth-century reference(area-weighted mean) -14 °C.

This points to using the global mean temperature to evaluate evaporation over the entire planet. This is the temperature, Ts, evaluated by our model. Investigating this further, NASA Science informs us that 86% of global evaporation occurs in the oceans. Using this to scale the mean temperatures above gives us an estimate of 15°C for an effective global **evaporative** temperature. This informative check is only one degree higher than the mid-twentieth century reference of about 14 °C.

Given the lack of rigour in the above estimate, an offset facility for ocean temperatures is included in the model. Use was made of this offset facility to perform a sensitivity check. Imposing a +2°C offset changed the modelled global temperature by 0.002°C in 2019 and 0.01°C in 2100. Even a 5°C offset gave a change in 2100 of 0.04°C. This latter offset is equivalent to a global mean ocean surface temperature of 21°C (unrealistically high). The bottom line is that the above approach to water vapour feedback is relatively insensitive to the absolute ocean temperature but appears dominantly sensitive to the changes in humidity induced by global warming temperature changes.

One further assumption in the analysis is that the temperature on top of the boundary layer is essentially the same as the temperature 1.5 metres higher. This appears consistent with similar assessments (by climate scientists) involving ground surface temperatures. Further, the calculations involved here work on temperature differences from one level of greenhouse gas to another.

The above points suggest that it is appropriate to use the overall global mean value of Ts (evaluated from our model) as the basis for calculating water vapour feedback.

Notes:

The value for F_{LH} is one of several values stated / updated by NASA. I have used a value of 86.4 W/m² for consistency with my initial work. The same applies to the value of 40W/m² for Winflux.

The source of the polynomial used for water vapour density is a George State University (Hyperphysics) website. It is appropriate up to 40 °C and was checked as numerically good in terms of differences between temperatures. This was selected because only four terms are needed – maintaining my approachable aim.

Whilst Ts, the surface air temperature, is used in the estimation of evaporation, it is important to realise that the sun heats the ocean, and the ocean heats the air. In this analysis, we rely on the measured temperature difference between ocean and air, but implicit is the heat transfer process noted above.

Further Information

- In this appendix, I have assumed that the air surface temperature immediately above the ocean is essentially the same as the temperature at a height of 1.5m. Figure 6a of the report below lends support to this approach.

    Air temperature profile and air/sea temperature difference measurements by infrared and microwave scanning radiometers. D. Cimini et al. 2003

Appendix 6 - Atmospheric Heat Gain Due to Increased Water Vapour

Change in Absorption of Sunlight

It is reasonable to assume that additional water vapour in the atmosphere would absorb more sunlight. It follows that as it is captured by the atmosphere, this captured sunlight would not directly heat the surface. It is also true that the heat captured by the planet in total would remain the same. So, this appears to be a zero-sum process. However, there is a subtlety buried in this analysis. That subtlety is that the two heat transfer processes (atmospheric transfer and window radiation) are different, and varying the proportion of heat transferred in each process will result in a different outcome. Dealing with this by reviewing the effect on the atmospheric transfer route is difficult and would involve considering heat capture at different altitudes (etc..) and its ultimate impact on surface temperature. I have chosen to make an estimate using the simplest approach.

Figure A6.1 shows planetary heat flows obtained by NASA from satellite observations. From this we can see that less heat captured by the surface would result in less heat radiated via the thermal atmospheric window. This is not an easy estimate to make directly. To resolve it I used the values in Figure 6.1 to work out the proportional change in window radiation. I also used our model to determine the initiating change in temperature caused by a change in CO_2 level (with evaporative feedback enabled). Using several different CO_2 changes, I calculated a feedback effect of $0.3W/m^2/°C$. This is a very rough and ready approach, yielding a small value, but for completeness I included it in the model.

Note: Figure A6.1 is from an era that aligns with other data used in Appendix 5

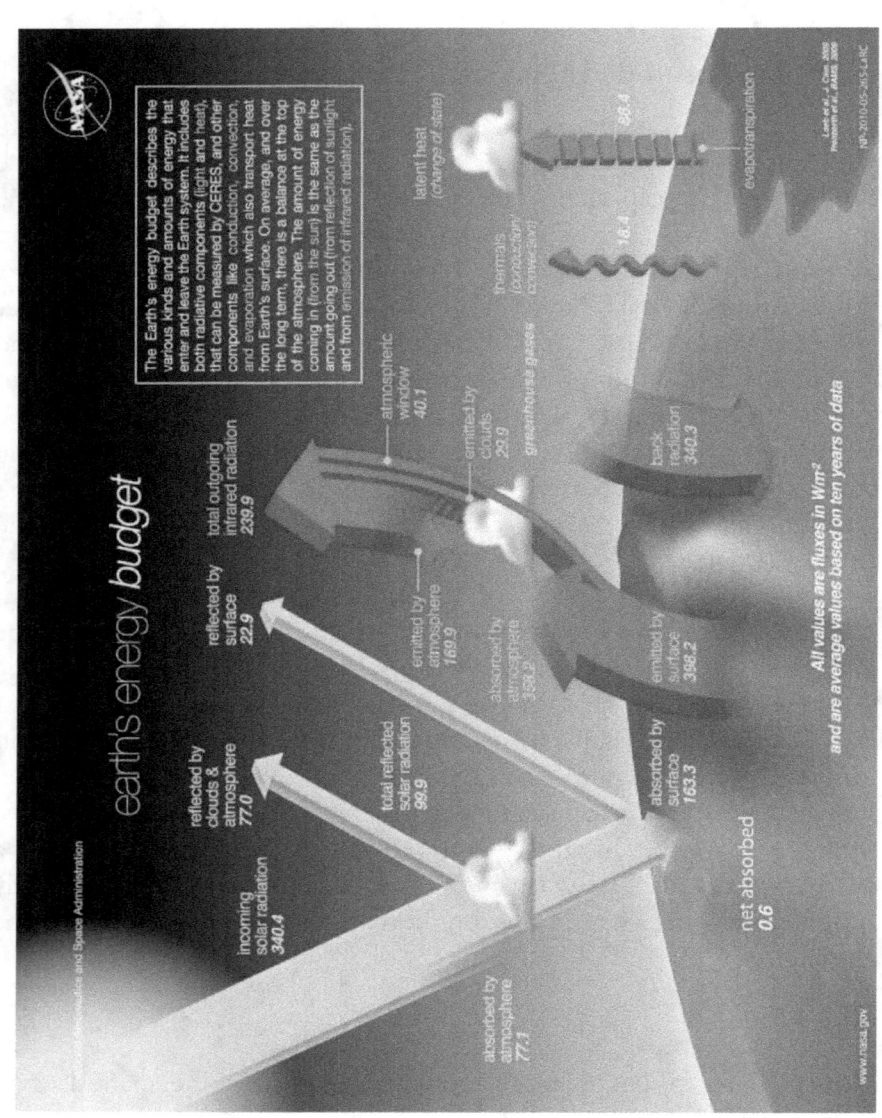

Figure A6.1 Earth's Heat Balance, NASA

Water Vapour Absorption Effects on Thermal Window Radiation

The change in mass of water vapour may also additionally increase the atmosphere's absorption of thermal radiation. This could in turn reduce radiation via the atmospheric window. As less heat would escape via this route, it would also heat the atmosphere.

However, in the thermal atmospheric window region, the majority of absorption by water vapour is already at 100%. This means that only the residual amount is subject to change.

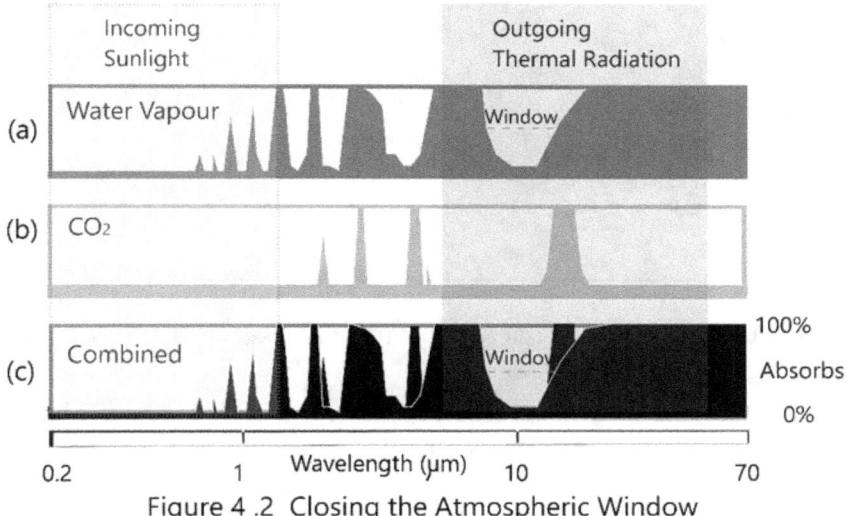

Figure 4.2 Closing the Atmospheric Window

This is shown in Figure 4.2, which is repeated here. Further to this, a more detailed plot would show that, in terms of the absorption coefficient, the low point in Figure 4.2 is about 10,000 times smaller than the 100% absorption level. More detail is shown in Figure A6.2 below, where the fuzzy line is the absorption plot for water vapour. From this, it seems the opportunity for increased absorption in the thermal window region would be small. For this model, I have chosen to ignore this effect.

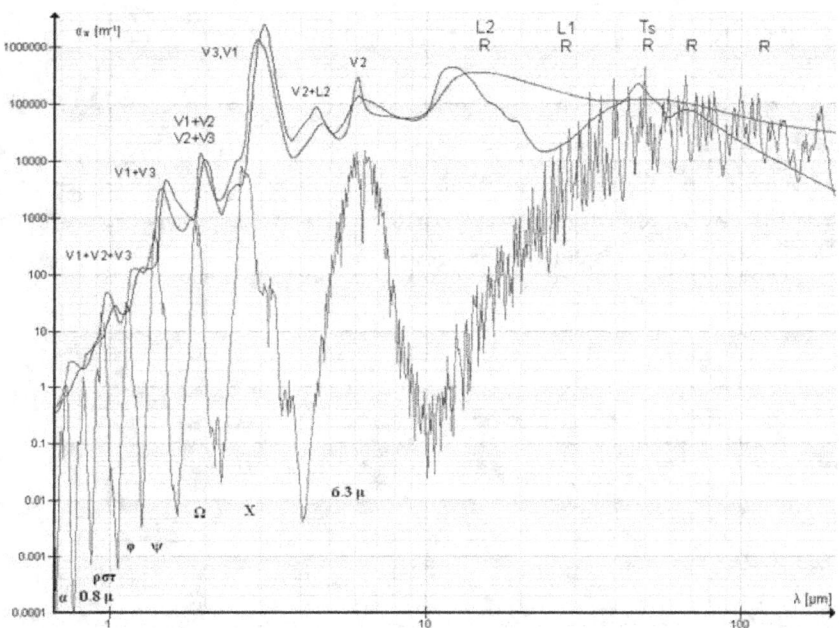

Figure A6.2 Electromagnetic Absorption by Water

Image produced by John Bertie and featured in Wikipedia. The CO_2 absorption band is located at about 15μm

Appendix 7 - Calculating Radiative Forcing

Radiative Forcing, or RF, is the instantaneous power input that the earth's climate system receives due to increases in greenhouse gasses or other changes. This change in power is assessed between conditions at the instant of change and those in 1750 (The pre-industrial period). Importantly, it is assessed at a point <u>before</u> the Earth's climate systems respond, and therefore, the Earth's surface temperature is assumed constant. RF has units of W/m^2.

Greenhouse Gas Radiative Forcing

The spreadsheet needs to calculate this to give us a Quality Assurance check when we make use of IPCC RF relationships and convert them to absorption values. It is also needed if we wish to inject other changes into the model where we only know the RF.

For our simple model, a calculation of RF is relatively simple. When we are considering greenhouse gasses, we are looking at changes in absorption by these gases. For absorption, we have a relationship from Section 4:

$$\text{Winflux} = \text{Tgh} \cdot \sigma \cdot \text{Ts}^4$$

Where Tgh is the transmission factor for heat radiated from the earth's surface. We can see two variables here, the transmission factor **Tgh**, and the surface temperature, **Ts**. The change in Winflux between 1750 and our date of interest is the Radiative Forcing. It is the change in power radiated as atmospheric transmission **Tgh** varies.

$$\text{RF} = \Delta\text{WinFlux} = \Delta\text{Tgh} \cdot \sigma \cdot \text{Ts}^4$$

and as Ts is assumed constant

$$\textbf{RF} = (\sigma \cdot \textbf{Ts}_{1750}{}^4) \times \Delta\textbf{Tgh}$$

Where $\textbf{Ts}_{1750}$ was the global average temperature in 1750, and **ΔTgh** is the change in transmission factor between 1750 and another date of interest. You may have noted that RF is the derivative of Winflux with respect to Tgh.

IPCC reports state a figure of 280ppm CO2 for 1750, and my estimate of the temperature is 286.76K. From these conditions, our model can evaluate RF for any CO_2 level (enabling comparisons with IPCC RF values.)

Solar Input Radiative Forcing

When we are looking at changes that affect solar input, these are assessed separately by means of a change in albedo. Therefore, the RF is the difference between the solar input power at the pre-industrial albedo compared with the power at some later time when the albedo is changed. Recalling Equation (1) from Section 3

$$\text{SolarPower} = Sc \times (1-\text{albedo}) \times \text{CaptureArea} \quad \ldots\ldots\ldots(\text{Watts})$$

$$RF = Sc \times (\Delta\text{albedo})/4 \quad \ldots\ldots\ldots(\text{Watts}/m^2)$$

The factor of '4' comes from the difference in area using a capture cross section compared to the curved surface of the globe.

Appendix 8 – Guide to the Spreadsheet

I have used an Excel spreadsheet because I am familiar with it; it is widely used and is installed on my laptop. If you do not have this software, I apologise and hope you will understand that providing a range of spreadsheets was not my objective.

The spreadsheet you can download is essentially the same one used to support the Global Warming Handbook and is essentially a 'workbook'. It uses only the most basic Microsoft Excel facilities and is saved in a file format called 'compatibility mode', which should enable its use with older versions of Excel. There are no macros or Visual Basic coding to trip you up – and I am hopeful that compatibility issues should be minimised. Also, I have run it on my tablet using Microsoft OneDrive.

Before exploring the spreadsheet, it is a good idea to check that iteration is enabled within Excel. This is enabled via the menu sequence File>Options>Formulas – set the maximum iterations to 100 and the maximum change to 0.001. This is set on my spreadsheet, but it can revert to the default for your Excel package. If you are prompted to make a backup copy, you should do so, or some spreadsheet functions may not work.

The spreadsheet, as downloaded, will have write protection on all worksheets. This is to protect you from accidental alteration. All cells where data entry is expected are unprotected. You can easily unprotect worksheets as no password is required.

The spreadsheet is protected by my copyright, and anyone is free to use it as an individual for non-commercial use (if they accept my disclaimer). That means you should not copy it for commercial reasons or sell or distribute it in part or in full. Please use it fully and provide attribution where it contributes to your use.

Using the Spreadsheet

The spreadsheet comprises a collection of worksheets, and you can select the one you want from the tabs at the bottom of the screen. Each worksheet deals with a different aspect of the model

To run the model and view plots, you only need to use two worksheets: **Control Panel** to choose reference conditions; and **Auto**

Chart to view the result with a 'dial-up' range of global warming processes to try out.

You may also be interested in the **CO2 and Window** worksheet as well as the **Warming** worksheet, which exposes the calculations explained in this book. However, there are also several other worksheets that contribute to the model or the background to the Global Warming Handbook.

Use is made of dropdown lists to configure the model. This makes it easy to run – just remember that the option cells are not push buttons - you must click on them to show the dropdown control and then click on the control to view a list of options (which you can finally select).

Each Worksheet within the spreadsheet is identified by a name displayed on its **Tab,** and each of these is briefly explained below:

Control Panel Tab

This enables you to set the reference condition for the worksheets that follow. A number of these are immediately defined and can be easily selected using a dropdown list. Additionally, there are several non-committed entries which you can load yourself and then select to use. No calculations were originally performed on this sheet. However, the CO2 % entry for 'All WMGHG 1890 Reference' is now updated automatically from the WMGHG Tab.

	CO_2 absorption - Available Reference Conditions - these apply to all Worksheets					
		Ts	CO_2	Thermal WinFlux	Relative Absorption	Albedo
		(K)	(ppm)	(W/m^2)	(CO_2 %)	
	Preset Conditions					
1	◯ Try 3.62% 1890 ref	286.80	292.46	42.48	3.620	0.3000
2	◯ Try 2.77% 1890 ref	286.80	292.46	41.85	2.770	0.3000
3	◯ IPCC CO2 Only 1890 Reference	286.80	292.46	41.10	1.530	0.3000
4	⦿ All WMGHG 1890 Reference	286.80	292.46	41.64	2.665	0.3000
	Initial Reference examples					
5	◯ IPCC CO2 only 2005 Ref	287.83	378.90	40.00	1.530	0.3000
6	◯ All WMGHG 2005 Ref	287.83	378.90	40.00	2.371	0.3000
	User Values					
7	◯ User Values 1					
8	◯ User Values 2					
9	◯ User Values 3					
	You can edit the User Values and then select them					

Importantly, the Control Panel Worksheet also contains a drop-down box labelled 'Iteration'. This switches the iterative nature of calculations in the worksheets that follow. The options available are 'On' or 'Reset'. Its main use is to recover when the model is driven to an unusual state. This may occur as you experiment and change or adapt the worksheet.

CO_2 and Window Tab

This worksheet uses the reference conditions set in the **Control Panel Tab** to work out the values of key model parameters. This includes constants for calculations relating to the thermal Atmospheric Window, Atmospheric Heat Transfer calculations as well as Albedo. This was illustrated in Figure 6.1.

The calculation of the reference parameters is shown explicitly so you can see each stage of the calculations. From this, you can see how the equations explained in the book are realised. If you amend this sheet, you are making fundamental changes to the model.

Warming Tab

This sheet performs an individual calculation of the mean annual global temperature. You can enter a value for CO_2 concentration and view the result. Again, each stage of the calculation is exposed. It is a good place to start and familiarise yourself with how the model works. Additionally, it provides an estimate of Radiative Forcing since 1750. This is calculated separately for changes that affect the Thermal Atmospheric Window and those that are induced by changes in albedo (solar input effects). Currently, this is the only place in the spreadsheet where this is evaluated explicitly.

No Greenhouse Tab

This is a stripped-down version of the **Warming Tab** with CO_2 and water vapour absent (set to zero concentrations). It is best done on this separate Tab and is primarily a cross-check.

Auto Chart Tab

Once you have set you reference conditions (**Control Panel Tab**), you may never have a reason to leave the **Auto Chart Tab**

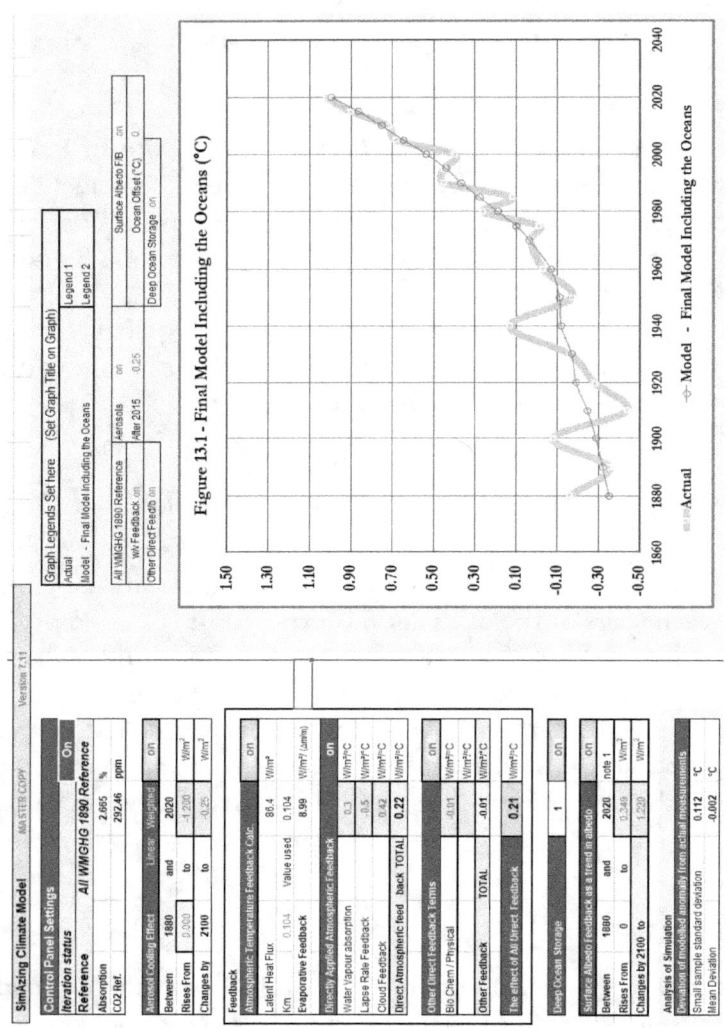

Figure A8.1 Auto Chart User Layout

Figure A8.1 shows the layout of the top left part of the worksheet. On the left-hand side are several control boxes, and to the right of them are a number of display charts.

Aerosol Cooling Effect		Linear	Weighted	on	
Between	1880	and	2020		
Rises From	0.000	to	-1.200	W/m²	
Changes by	2100	to	-0.25	W/m²	

Feedback			
Atmospheric Temperature Feedback Calc.		on	
Latent Heat Flux		86.4	W/m²
Km	0.104 Value used	0.104	
Evaporative Feedback		8.99	W/m²/ (Δm/m)
Directly Applied Atmospheric Feedback		on	
Water Vapour absorption		0.3	W/m²/°C
Lapse Rate Feedback		-0.5	W/m²/°C
Cloud Feedback		0.42	W/m²/°C
Direct Atmospheric feedback TOTAL		0.22	W/m²/°C
Other Direct Feedback Terms		on	
Bio Chem / Physical		-0.01	W/m²/°C
			W/m²/°C
Other Feedback TOTAL		-0.01	W/m²/°C
The effect of All Direct Feedback		0.21	W/m²/°C

Deep Ocean Storage	1	on

Surface Albedo Feedback as a trend in albedo			on	
Between	1880	and	2020	
Rises From	0	to	0.350	W/m²
Changes by 2100 to			1.237	W/m²

Figure A8.2 Auto Chart Control Boxes

Using these boxes, you can introduce the effects of aerosols or feedback and instantly see the plot of the outcome. The charts run in sequence across the page, covering a range of dates, both for contemporary dates and future dates. Whilst Figure A8.1 is a general view and on a small scale, the details of the control boxes are more clearly shown in Figure A8.2.

You can set conditions that are described in the book using drop-down ' list boxes '. These have a blue background and red text. Simply click on them to access the ' drop-down list ' Using them, you can turn on aerosol effects and feedback effects. Note that the absorption factor for greenhouse gasses cannot be amended here but needs to be initially set using the **Control Panel Tab**.

In addition to using the pre-set conditions called up using list boxes You can also amend the parameter values called up by these list boxes. These are pre-set to values explained in this book. These are identified by red text on a white background. Clearly you will need to exercise some care here – remember the 'Iteration Reset' function available in the **Control Panel Tab** (you may need it!) – as well as your backup copy of the spreadsheet.

With the spreadsheet as downloaded, you need to alter nothing to match the presentation in the series of books.

Auto Chart Tab - Model Implementation:

If you looked at the **Warming Tab** you will have seen the modelling calculations exposed for an individual calculation of global mean temperature for a given CO_2 value. Now, when you look at the **Auto Chart Tab** you will see, below the graphs, that this calculation is made as a single line. Each line starts with a date and a historical CO_2 level. Starting with an estimate of the surface temperature, it progresses along the line, working out thermal Atmospheric Window Radiation, Atmospheric Transport and then using Solar Input it works out an amendment to the surface temperature. this amendment is passed back to the initial estimate. This is done iteratively until the changes to surface temperature are small (see Section 6). Further, as the line progresses, further, changes due to albedo (aerosols) atmospheric feedback (using a calculation of saturated vapour density at the ocean surface)., and ocean heat storage are also included For Atmospheric Feedback, you can make an offset adjustment to sea surface temperature, if you wish, using one of the control boxes in the top left of the spreadsheet. Due to the configuration of the model, this is only meaningful if Deep Ocean storage is disabled

In total 20 lines cover 20 dates and CO_2 levels from 1880 to 2020. A further three lines cover future predictions.

Returning to the upper left of the worksheet, you will find a simplified assessment of the model performance with the settings you have chosen

Analysis of Simulation

Deviation of modelled anomaly from actual measurements		
Small sample standard deviation	**0.112**	**°C**
Mean Deviation	**0.004**	**°C**

Run-time temperature and Feedback Assessment			
TCR estimated for 2 x CO2		2.04	°C
ΔTs (°C) without feedback/aerosols		0.953	°C @2020
Feedback	1.040	1.092 (W/m²/°C)	@2020

Water Vapour Change by	2020	8.7%
Water Vapour Change by	2100	26.8%

The TCR calculation is the temperature rise for an instantaneous doubling of CO_2. To avoid undue complication, a change in greenhouse gasses that equates to a doubling of CO_2 (between 2015 and 2077) is used. The forcing for a doubling of CO_2 can be evaluated using the **Warming Tab** for a CO2-only setting on the Control Panel Tab (a forcing of 3.63 Wm^{-2}).

A display of the total feedback used in the simulation (between 1890 and 2020) is also displayed. This needs the 'modelled value' of the temperature anomaly in 2020 to be provided (ΔTs) with all feedback and aerosol options disabled

The model is not set up to calculate more detailed feedback terms without error-prone tinkering and isolating one change from another. I would recommend avoiding this as the outcome can be misleading.

MWGHG Tab

This Tab exposes the data used to determine the effect of greenhouse gasses other than CO_2. For convenience, the value of greenhouse gas forcing as a percentage of all greenhouse gasses is fed back directly to the **Control Panel Tab** setting for the pre-set conditions for 'All WMGHG 1890 Reference'. In addition, it contains an option to include or disable the Forcing effect of Ozone.

Aerosols and Albedo Tab

This calculates the time-related ramp of albedo adjustments. To do this, it takes the conditions from the 'Aerosols and Cooling' panel on the **Auto Chart Tab** and translates them to an albedo ramp. This avoids over complication of the **Auto Chart Tab**. The outcome is fed back for use in the albedo adjustment section on the Auto Chart Tab.

It performs a similar function to translate the settings on the 'Albedo Feedback Panel' of the **Auto Chart Tab**.

Ocean and Imbalance Tab

This evaluates the heat transfer parameters associated with the energy budget imbalance highlighted in AR6. This is used to set the feedback used in the model to allow for thermal storage in the ocean depths.

You can adjust Tref to obtain the same heat transfer value under preindustrial and current conditions.

The worksheet evaluates and displays Tref, and you manually enter this into the heat balance calculation. This has been done this way as manual adjustment of Tref could enable the use of different heat transfer terms for pre-industrial and contemporary conditions.

This Tab also provides the justification for the use of a feedback term based upon the long thermal time constant of the ocean depths.

Surface Areas Tab

This Tab is for information only and exposes the data used to calculate global surface areas used in the model.

Thermal Masses Tab

This Tab is for information only. It exposes data used to evaluate the thermal masses of land, sea, and atmosphere used in the **Ocean and Imbalance Tab** as well as in the Global Warming Handbook).

Forcing Tab

This worksheet is used by the **Warming Tab**. It performs the estimation of Radiative Forcing for given inputs of CO_2 or Albedo.

NASA GIS & CO2 Tab

This Tab is for information only and shows the historical record of temperatures and CO2 levels on a year-by-year basis, as well as their sources. It covers annual data from 1880 to 2020.

8.021P

www.ingramcontent.com/pod-product-compliance
Lightning Source LLC
Chambersburg PA
CBHW052325220526
45472CB00001B/269